# MARINE INVERTEBRATES
## A FIELD GUIDE FOR CHILDREN

*The <u>I Saw It!</u> Series #6: Field Guides, Journals, and Coloring Books for Children*

by

Barbara L. Brovelli-Moon

Illustrations by Amy V. Mitchell

**Junior Naturalist**_____ **Date**_____

ANCHORAGE, ALASKA

## *Acknowledgements*

Endless thanks go to so many people at the Alaska SeaLife Center in Seward, Alaska.

Special thanks to Richard Hocking, curator. Your patience and knowledge helped me with myriad questions. Thank you for always being available to explain what I didn't understand, to correct inaccuracies in "first writes," and to offer support and encouragement. Thank you.

Darin Trobaugh, huge thanks for compiling the list of inverts for this book. To select 33 from thousands . . . no more needs to be said except I am impressed and grateful!

Deb Magruder, thanks for your support, insights, ideas, expertise, and willingness to read pages as this project wandered its strange path. You're a great friend.

Nancy Anderson, thanks for always having a smile, hug, and laugh whenever I saw you on my roamings through the Center.

SLC aquarium staff, interns, and volunteers. From people at the reception desk to the touch tank, my daily visits to see, and often touch, the "critter of the day" were always met with welcome and interest. Thank you for sharing your knowledge, time, and enthusiasm about the animals as well as this project.

Norma Neill, friend, editor, and cheerleader. Thanks for all your time, grammar expertise, and overall English-language knowledge. Your willingness to read, reread, and read again amazes me. You know I can't do this without you . . . for sure!

Amy Frackman, tech wizard. The dead computer, no color with my printer, etc., etc., etc. Somehow you always know "the fix" so I keep breathing. Enchiladas whenever you want them.

Amy V. Mitchell, illustrator extraordinaire. So glad we connected and you jumped right into this project. The illustrations are amazing and exactly what I wanted. Thanks for sharing your talent.

Granddaughter Elli, your coloring of the octopus on the back cover is creative and so artistic. Thank you for doing that wonderful art.

And Greg, thank you for always being there as our life rolls along. I'm so grateful for you.

Special acknowledgment and thanks to the SeaLife Center for granting permission to use the cover photograph of an amazing octopus.

Text and illustrations copyright © 2020 by Barbara L. Brovelli-Moon

All rights reserved. No part of this book may be reproduced, scanned, or distributed in any printed or electronic form without written permission from the publisher.
Published by OceanOtter Publishing.
Printed in the United States of America. Designed by Phillip Gessert of Gessert Books.
Brovelli-Moon, Barbara L. (Illustrator, Amy V. Mitchell)
Marine Invertebrates: A Field Guide for Children.
The I Saw It! Series #6: Field Guides, Journals, and Coloring Books for Children
ISBN: 978-0-996-7-4
10 9 8 7 6 5 4 3 2 1
Contact the author at oceanotterpublishing@gmail.com or www.oceanotterpublishing.com

# Table of Contents
# And Checklist of Sightings

**Tidal Zones** ..................................... 2

**Introduction** ................................... 3

**Phyla** ............................................... 4

## Cnidarians

*Anemones*

❏ Burrowing Green Anemone ........................ 6

❏ Painted Anemone ................................ 8

❏ Plumose Anemone ............................... 10

*Jellies*

❏ Moon Jelly ..................................... 12

❏ Lion's Mane Jelly .............................. 14

## Annelids

*Polychaetes*

❏ Bristle Worms .................................. 16

## Arthropods

*Barnacles*

❏ Common Acorn Barnacle .......................... 18

*Isopods*

❏ Seaweed Isopod ................................. 20

*Amphipods*

❏ Common Beach Hopper/Sand Flea .................. 22

*Shrimps*

❏ Spot Shrimp .................................... 24

*Crabs*

❏ Hairy Hermit Crab .............................. 26

❏ Dungeness Crab ................................. 28

❏ Pygmy Rock Crab ................................ 30

❏ Graceful Decorator Crab ........................ 32

## Echinoderms

*Sea Stars*

❏ Mottled/True Sea Star .......................... 34

❏ Sunflower Star and Morning Sun Star ............ 36

❏ Blood Star ..................................... 38

❏ Ochre Star ..................................... 40

❏ Leather Star ................................... 42

*Brittle Stars*

❏ Daisy Brittle Star ............................. 44

*Urchins*

❏ Red Sea Urchin ................................. 46

❏ Green Sea Urchin ............................... 48

*Sea Cucumbers*

❏ California Sea Cucumber ........................ 50

## Mollusks

*Snails*

❏ Hairy Triton ................................... 52

❏ Sitka Periwinkle ............................... 54

*Nudibranchs*

❏ Opalescent Nudibranch .......................... 56

*Limpets*

❏ White Cap Limpet ............................... 58

*Octopuses*

❏ Giant Pacific Octopus .......................... 60

*Chitons*

❏ Black Katy/Black Leather Chiton ................ 62

❏ Lined Chiton ................................... 64

*Clams*

❏ Pacific Littleneck Clam ........................ 66

❏ Pacific Razor Clam ............................. 68

*Mussels*

❏ Pacific Blue Mussel ............................ 70

**Why Do We All Have Three Names?** ... 72

**Tide Pool Etiquette** ......................... 73

**So You Want to Be A Marine Biologist?** 74

**Mr. Richard and the Alaska SeaLife Center** ................................. 75

**Glossary** ........................................ 76

**Amazing Oceans Map** ..................... 78

# Introduction

What's *your* favorite animal? A dog? Cat? Horse? Pig? Dolphin? Penguin? Maybe you love birds, a certain fish, or even toads. These are all amazing, interesting animals.

And they are all vertebrates (VER-tuh-brayts), animals that have a backbone. We humans have a backbone, too. Feel the bumpy bones down your back. Those are called vertebrae (VER-tuh-bray). Our thirty-three small, ring-shaped bones connect to each other to make our spines. Our bodies also have other skeletal bones that work together to hold us in a particular shape, to help us move, as well as to help our bodies in other important ways.

Now, think about this. There are more than 1,250,000 (one million, two hundred fifty thousand) identified species of animals on our planet. Scientists say there are probably *millions* more we don't even know about yet! And, no more than five percent, maybe 60,000 (sixty thousand) species, are vertebrates with a backbone and skeleton. What about all the other 1,190,000 (one million, one hundred ninety thousand) animals? What *are* they? *Where* are they? What have I *missed*?

Well, all those other animals are *in*vertebrates, meaning they have *no* backbone. No backbone is the *only* characteristic common to all of them. These animals are the most widely varied, unique creatures on Earth. They live in the air, on land, in freshwater, and at all levels of all seas and oceans. Invertebrates come in all sizes, from the tiniest you see only by looking though a microscope, to the largest colossal squid that weighs over 1,000 pounds. Some live alone, others in small groups, and still others in colonies of thousands. Some eat plants, some eat "meat," and others eat everything. How do we even *begin* learning about all these unique animals?

We begin by thinking of groups. You know that scientists have classified all living organisms into groups to keep track of the exact animal they are studying, right? (Read more about scientific classification on page 72, "Why Do We All Have Three Names?) Well, all vertebrates *and* invertebrates are in the Kingdom *Animalia* (an-uh-MAIL-yah). The next division is Phylum (FYE-luhm), and that's where *our* study of invertebrates begins.

This book introduces you to thirty-three *marine* invertebrates from five of the many invertebrate phyla. *All* marine invertebrates live in, or right along the edge of, an ocean. (The Tidal Zones page to the left shows you these zones and a few of the animals who live in each one.) These unique and varied animals tell you all about themselves and their fascinating worlds along the Pacific coast of North America. By reading their stories and learning more about these amazing creatures, you'll surely become *curious* and want to learn even *more*. And *that*, my readers, is when even *more* fun begins . . . when *you* do your *own* research, writing, questioning, and learning about the incredible world of invertebrates. Enjoy!

# CNIDARIANS
(nye-DARE-ee-uhns)

From Greek word for "stinging nettle"

Simple animals who live in water. Most are marine, living in oceans and seas rather than freshwater.

All have stinging cells, tentacles, and a single mouth opening.

Sea anemones, jellies, corals, sea pens. At least 10,000 species of them!

Either bell-shaped and mobile, or have tubes attached to one spot.

Symmetrical bodies, so if cut down the middle, each half looks exactly like the other.

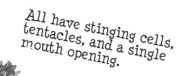

# ARTHROPODS
(AR-thruh-pahdz)

Greek for "jointed foot"

Largest and most diverse animal phylum with 85% of all known animals. Almost 1 million species with 90% being land insects.

All have legs, limbs, or feet with joints.

Ten percent, about 85,000 species, are marine animals living in oceans and seas.

All have segmented bodies and two pair of antennae. Most have well-defined heads.

All have hard exoskeletons. These are shed and regrown as animal grows.

Barnacles, isopods, shrimps, crabs, spiders, insects, ticks, and many more!

# ANNELIDS
(AN-uh-lidz)

Latin for "little ring"

Worms with bodies made of small, ring-like segments.

Polychaetes (PAH-lee-keetz): a group of bristly-worm annelids who live in all oceans of the world. Over 10,000 species of these worms!

Over 17,000 species all together, including earthworms and leeches.

No legs nor hard skeleton. Have leg-like paddles on each side of each segment.

Have head with mouth on underside, and eyes and antennae on top.

Some roam freely; others make permanent tubes in which they live.

# ECHINODERMS

(eh-KI-nuh-durms)

*Latin for "spiny skin"*

Body parts usually divided into multiples of five.

ALL have spiny or bumpy skin. No brain, heart, nor eyes.

Sea stars, brittle stars, sea urchins, sand dollars, sea cucumbers.

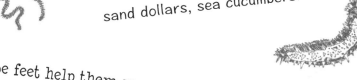

Tube feet help them move around *in* the water.

Water-vascular system moves water around *inside* the body.

Largest phylum that lives **only** in oceans. Live at all depths of all oceans, mostly crawling on the ocean floors.

Most regrow body parts, so they live long lives.

About 7,000 living species and 13,000 extinct ones.

# Mollusks

(MOL-uhsks)

*From Latin mollis for "soft"*

Second largest phylum with over 100,000 species. Largest marine phylum. Eighty percent of mollusks are gastropods.

Most live only in oceans; about 20% live on land and/or in freshwater.

Phylum with the most extreme variety of animals.

Sea snails and slugs, limpets, chitons, clams, mussels, octopuses.

Bodies are soft with no legs nor bones; many grow hard shells for protection.

All have …
   … a fold of the body called a mantle that holds their internal organs
   … a foot to help them move around
   … a sharp, tongue-like radula.

# Burrowing Green Anemone

*qaamida-n* in Aleut • *Anthopleura artemisia*

Oh, hello! I'm *so* excited to be the first animal in this book about the amazing underwater world of marine invertebrates! I'm Annabelle Anemone (uh-NEM-uh-nee), a sea anemone who digs down deep into the sand. In my drawing, you don't see my round, tall body column. I burrowed into the sand by slowly tightening and loosening the muscles along my soup-can-like column. I don't move down quickly, but once I'm buried, the round, sticky foot at the bottom of my body attaches firmly to a rock or shell deep under the sand. See why I'm called a buried or beach sand anemone?

In my drawing you *do* see the long, slender, arm-like tentacles that grow all around the upper part of my column. Those tentacles are green with white bands. The green is from the green algae that live in my tentacles. On the very top of me is my flat, oral disc with small tentacles surrounding my slit-shaped mouth. Those little tentacles are *extremely* dangerous, deadly for many tiny ocean creatures I eat.

You see, animals in my Cnidaria (nye-DARE-ee-ah) phylum have tiny nematocysts (neh-MEH-tuh-sists) in our tentacles. Nematocysts are thin, wound-up threads with barbed, or hooked tips. When a tiny fish or other tasty prey swims too close, my threads shoot out like harpoons and wrap around the victim. The barbed tips are needle-sharp, so they poke into my prey, stick, and shoot poison into it. Then, I pull the threads and food back into my mouth.

My mouth is the only opening in my entire body. That's where I take in food, and I also poop out my body's waste products there. (I know. Sounds yucky!) I drink seawater in through my mouth, too, and that's *very* important. I'm an invertebrate, meaning I have no backbone. The water I drink becomes like a skeleton for me. Water moves throughout my body giving my muscles something to push against. Without my water skeleton, I'd be a limp, falling-over, green blob with no shape at all!

Sometimes I *do* look like a sand-covered blob! I hide from my enemies by folding my column all the way inside, curling into a tight ball. Then I'm a still, small puckered hole in the sand. Look carefully on beaches of sheltered bays, in tide pools, and along open shores. You *might* see me!

## My Facts

**SIZE**: Column height: up to 10 inches. Crown/top diameter: up to 4 inches. **COLOR**: Oral disc: bright pink, orange, or green. Upper column: greenish or brownish on exposed part; black, gray, brown; Lower column: white or pink. Tentacles: green, solid or banded with white; red, white, black, blue gray, brown, pink, or orange. **FOOD**: Carnivorous. Small fishes, crustaceans, various invertebrates, spawned herring eggs, zooplankton. **PREDATORS**: Nudibranchs, small fishes, sea stars. **TIDAL ZONE**: Low intertidal, shallow subtidal to 98 feet deep. **RANGE**: Gulf of Alaska to S. California. (Very common around Juneau, Alaska.) **LIFESPAN**: up to 50 years.

## Did You See Me? Tell Your Story! _____

_____

_____

**DID YOU KNOW?** When you see me in a tide pool, you might want to poke me. *Please! Gently* touch my outer tentacles *only*. Yes, I'll squirt water and fold inward. But I need that water to stay hydrated at low tide when the sun's hot. Be thoughtful!

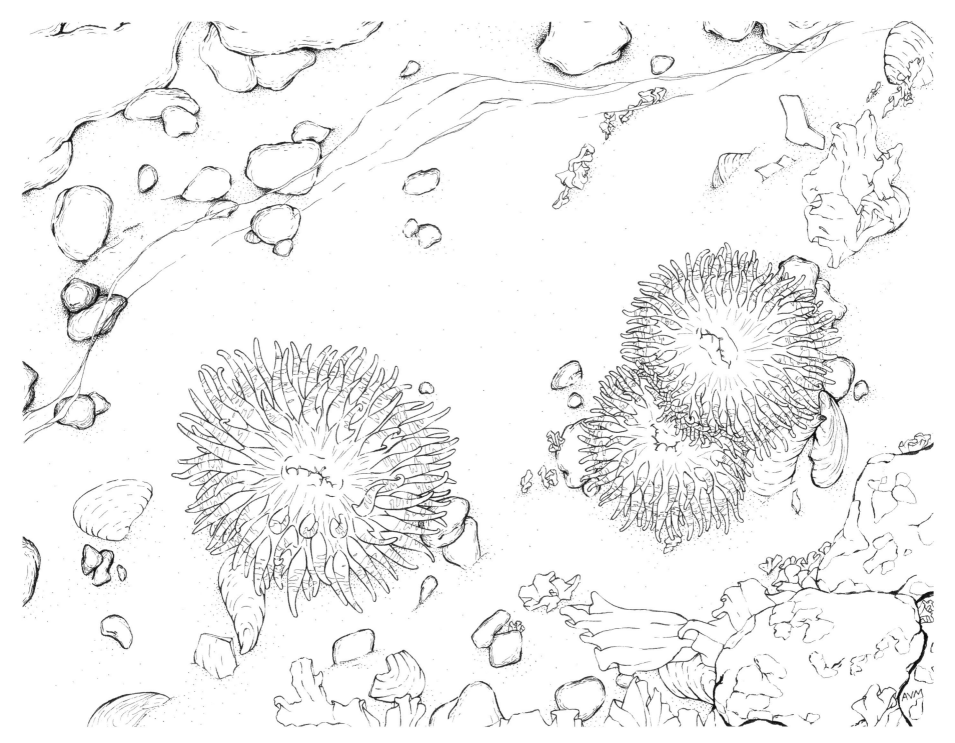

*Burrowing Green Anemone*

# Painted Anemone

*tsanagha* in Kenai Tanaina • *Urticina grebelnyi*

Greetings from my favorite tide pool! I'm Andy Anemone (uh-NEM-uh-nee), a very colorful and *splotchy* painted anemone. Doesn't it look like an artist filled a brush with red and green paint, then stood back and just threw globs of it all over my bumpy body column? Plus, my many long, pale, flowing tentacles are covered with pinkish-red, lilac, or brownish bands. I even have a thin, red band where each tentacle connects with the round oral disc on the top of my body. When danger is near, I try to hide by pulling all my colorful tentacles inside my column. Then I'm just a brightly painted ball on a big, flat foot like Amy on the rock over there.

Being so colorful led to my being named Mottled Anemone, Red and Green Anemone, and Christmas Anemone. I should be called *Bigfoot* Anemone because my pedal disc, my foot, is huge! It's much wider than my body column and oral disc, and it's really sticky on the bottom. That makes it easy for me to stay attached to large boulders or rocks, under ledges, or in crevices. I can't attach to sand . . . it's too loose. So don't look for me where Annabelle Anemone lives and burrows. Sand on my foot or column totally bugs me. Ugh!

Once I attach my foot, I stay put. In fact, I stick so tightly that if you try to pull me loose, you'll harm me! I *do* move when I want to be in an ocean current with more food, or if I need to get away from a dangerous creature who's wandered into my neighborhood. When I *do* move, I slowly loosen my foot, pull away from the hard surface, then glide or creep *slowly* along on my pedal disc. Don't bother trying to *see* me move. That would be like watching paint dry or grass grow. It's a very slow, boring process.

Since I *am* slow, it's a good thing I have built-in ways to protect myself and to capture food. Like all the other Cnidarians (nye-DARE-ee-uhns), hidden in my long tentacles are stinging nematocysts (neh-MEH-tuh-sists). When an enemy comes too close, the harpoon-like threads shoot out, the poison in the barbs stuns my enemy, and I have time to make my slow escape. I capture food particles the same way . . . shoot, stun, and paralyze. Then it's easy to pull the food back to my mouth. I may be slow, but don't worry. I survive quite nicely.

## My Facts

**SIZE**: Column height: 8 - 20 inches. Crown/top diameter: 6 - 10 inches. Pedal/base always larger than crown/top. Tentacles: up to 1.5 - 2 inches long; 0.4 inches diameter at base. **COLOR**: Oral disc: pale yellow green, lilac, or brownish; short, thin, red bands where tentacles attach to column. Column: varies; large, irregularly-shaped red and green patches, some more red or more green; often green to light yellow patches with red, blood-like markings; tan, red-green, or olive stripes. Tentacles: see story. **FOOD**: Carnivorous. Small fishes, various invertebrates, shrimps, krill, mussels, anything organic that drifts close enough to catch. **PREDATORS**: Nudibranchs, sea stars, snails. **TIDAL ZONE**: Low intertidal to 98 feet deep. **RANGE**: Gulf of Alaska to S. California, especially in N. Pacific. **LIFESPAN**: 60 - 80 years.

**Did You See Me? Tell Your Story!** _____
_____
_____

**DID YOU KNOW?** I'm a very social creature, so I'll probably be with a small group or even a large colony. Together, we form brightly colored carpets all over the ocean floor. Watch our tentacles wave gently at you when ocean currents ripple us.

Painted Anemone

# Plumose Anemone

*tayataayi'* in Tlingit • *Metridium farcimen*

Hey! Down here in front! It's little me, Avalyn, a plumose (PLU-mose) anemone. See the white, feathery plumes on top of my smooth, slender column? That's why I'm a plumose, 'cuz it means "having plumes." These beautiful feathers make me *so* easy to recognize. My kind are the *largest* anemones in the whole northern Pacific Ocean. Some of us grow to be *forty inches* when we stretch up tall. Hope *I* grow that big!

Look on rocks and dock pilings, and around harbors, jetties, and wharves. You'll probably see us! You *might* see just one plumose, but you'll probably see a colony, or group, of us. I was born into this colony when an older plumose pulled away from the rock where it was attached. That elder left behind a small piece of its pedal disc, or foot. That tiny piece grew slowly into a new anemone . . . *ME*! I'm a clone, or *exact* copy, of the elder. In fact, *all* anemones in my colony are clones of each other. It's strange that we aren't all the same color and size, but we all have *feathers*!

Our plumes are actually hundreds and hundreds of short, slender tentacles. These grow all over lobes, small, raised, rounded bumps, that are around the mouth on our oral discs. *Like* other anemones, these tentacles shoot poisonous nematocysts (neh-MEH-tuh-sists) into our prey. *Unlike* other anemones, we have something even *more* dangerous. Some of our elders stay on the edges of our colony. They have up to *nineteen* thick, long "catch tentacles" growing right on their lips. Inside is a unique, poisonous, stinging nematocyst whose tip goes into the enemy, breaks off, and *stays* stuck. That tip kills the part of the victim where it's stuck, or even kills the *whole* creature! When anyone who isn't a clone of our colony comes too close, one of our elders protects us by shooting their special weapons.

Isn't that clever, *especially* for the simple animals we are? Anemones have very few types of cells, those itty-bitty bits that make up all living things. You see our *outside* layer of cells. *Inside*, we have cells that work together to help us digest food then poop out any waste. Other cells make glue-like jelly that holds our outsides and insides together. We have no brain, no heart, no lungs . . . *nothing*. But don't forget our *fantastic* feathers! I just wish mine were *pink*!

## My Facts

**SIZE**: Column height: 20 - 39 inches. Crown/top diameter: up to 10+ inches. Tentacles: feeding, up to 4 inches; catch, up to 4.7 inches. **COLOR**: Oral disc/column: usually white; can be cream, orange, tan, brown, salmon, greenish. Tentacles: usually white; color of column when contracted; transparent when extended. **FOOD**: Carnivorous. Small invertebrates, zooplankton, other small, drifting food particles. **PREDATORS**: Nudibranchs, sea stars. **TIDAL ZONE**: Low intertidal/subtidal to 650+ feet deep. **RANGE**: Gulf of Alaska to S. California, especially in N. Pacific. **LIFESPAN**: 50+ years.

## Did You See Me? Tell Your Story!

_____

_____

_____

**DID YOU KNOW?** Sometimes I fold *all* my tentacles inside my column, and then I look like a ball on a flat disc. This protects me from predators. Folding in also helps me conserve water when the tide goes out, and I'm left in the hot sun!

Plumose Anemone 11

# Moon Jelly

*qaacek* in Kenai Alutiiq • *Aurelia labiata*

Jeffy Jelly here, and you *need* to know . . . jellies are *NOT* fish. I'm called a "jellyfish," but I don't have bones, gills, blood, brains, heart, nor lungs. Fish have all those parts. Jellies? We are "gelatinous zooplankton." That's "je-LAT-i-nes ZO-uh-plank-tn." I'm made of a jelly-like material, and I drift in water. Jellies have been drifting throughout oceans for more than seven hundred million years, *millions* of years before any dinosaur lived on Earth.

I'm a pretty amazing animal. See my drawing? That's me as a fully-developed medusa (muh-DUE-sah). I'm made up of about ninety-five percent water and five percent solid. The umbrella-like top is my bell, wider than it is high, with an outer and inner layer. Four, frilly oral arms hang down from the underside of my umbrella. These help me catch food then move it to my mouth, centered at the top of these arms. See the four, colorful horseshoe-shaped gonads on the outside of my bell? Jelly eggs begin growing in these. Also, on my outer layer, you'll see tiny, curvy lines. These are connecting, road-like canals, or tubes, that carry food, water, and waste throughout my body. Finally, on my bell's margin, or outside edge, hundreds of short, fringe-like, stinging tentacles help me capture food. Amazing, right?

Like jellies have done for millions of years, I float where ocean currents take me. I have *some* control in moving up and down, or on a slant. My bell pulses, opening when I relax my muscles, then closing when I tighten them. That makes me push against the water, thrusting me forward. It's like the push you get doing the frog kick when you swim.

I didn't *always* float or swim. At first, my tiny egg hatched into a baby larva called a planula (PLAN-u-la). Planula-me sunk to the ocean's floor and attached to a rock. There, I slowly grew into a polyp (PAH-luhp), a little bump that looked like a miniature anemone. Over time, I lost my tiny tentacles *and* divided into exact copies of myself through a process called strobilation (STRO-buh-lay-shun). My strobilae (STRO-buh-lee), or copies, stacked up like tiny, pie-plate discs. Each tiny disc peeled off the top of the stack and floated away. Now called an ephyra (eh-PHY-rah), every little copy of me grew into a small medusa. A new jelly! Aren't I completely amazing?

## My Facts

**SIZE**: Bell diameter: up to 15 inches. Tentacles: up to 4 inches. **COLOR**: Bell: juveniles/young adults, translucent; mature, milky white; depending on diet, can have pink, lavender, orange, peach, or blue tint. Tentacles: translucent white. **FOOD**: Carnivorous. Tiny zooplankton, mollusks, crustaceans, larvae of fishes and/or other marine animals, other small jellies. **PREDATORS**: Humans, other jellies, large fishes, sea turtles. **TIDAL ZONE**: Low intertidal to 3,200 feet deep. **RANGE**: Coastal waters around the world; S. Alaska to S. California; also common in N. Pacific across to Japan. **LIFESPAN**: A few days - 2 years.

## Did You See Me? Tell Your Story! _____

_____

_____

**DID YOU KNOW?** A group of jellies is called a "smack," and millions of us swarming together is a "bloom." Look for us floating near the ocean's surface or close to shore in bays and harbors. You may even see us washed up on beaches!

Moon Jelly    13

# Lion's Mane Jelly

ts'ahwa'daangaa in Haida • *Cyanea capillata*

Don't I have the most beautiful "hair?" I'm Jelena Jelly, named Lion's Mane Jelly because of my long, thick, mane-like tentacles. I'm among Earth's most beautiful creatures! I *am* the world's largest jelly and *may* even be the world's largest animal. Blue whales grow to be one hundred ten feet long, *supposedly* being the largest. Well, my kind has tentacles as long as one hundred *twenty* feet! That's longer! And beware! Our long tentacles are *extremely* dangerous to creatures in the sea and on shore, including *you*! Read on *very* carefully!

I have a smooth, flattened bell, thicker in the middle and thinner toward the edge. It's divided into eight lobes, or sections, with "innies" between the lobes. (I actually look like an eight-pointed star when I'm swimming and have pushed hard through the water.) On the underside of each lobe, close to my mouth, long, thin tentacles hang in sets of four rows. Each set holds from seventy to one hundred fifty tentacles. That means I have as many as *twelve hundred* tentacles spread all around my bell!

Remember: Cnidarians (nye-DARE-ee-ahns) all have tentacles with poisonous nematocysts (neh-MEH-tuh-sists). When I hunt for food, I push *my* tentacles down and outward. They spread like a net to make a circle around the fish or other prey I want to catch. As my victim brushes against the nematocysts, it's stung and stunned, an easy meal for me.

Now, this is important! If *you* brush up against Jeffy Moon Jelly, or most other jellies, you *migh*t develop a slight rash for a few hours. Anemones? Nothing happens if you touch most of them, or any other Cnidarian for that matter. But *DON'T* touch me! My poison is *extremely* strong, even for humans! A sting from me may cause a rash or searing pain for a few *hours*. You may react with severe burns and blistering, even *death*! A dead one of us who's washed up on shore is *still* dangerous, with a "sting" capable of making you *very* sick. I don't purposefully hunt you, but I sting more people in Alaska than any other jelly.

So be aware! We come in different sizes and colors. Young and small, we are yellowish or tannish orange. As we grow older and larger, we change into a deep brick red or purple, with reddish or yellowish tentacles. Remember: We are *DANGEROUS* at any size or color!

## My Facts

**SIZE**: Bell diameter: 1.5 feet - 6.5 feet; up to 7.5 feet. Tentacles: 30 feet - 120 feet. **COLOR**: See story. **FOOD**: Carnivorous. Anything small that touches tentacles; zooplankton, small fishes, crustaceans, nudibranchs, other jellies. **PREDATORS**: Leatherback sea turtles, anemones, large fishes, other jellies, sunflower sea stars. **TIDAL ZONE**: Intertidal to 66 feet deep. **RANGE**: Arctic Ocean, N. Pacific to Washington, sometimes Oregon, across Pacific to Asia; N. Atlantic. **LIFESPAN**: +/- 1 year.

## Did You See Me? Tell Your Story! _____
_____
_____

**DID YOU KNOW?** I'm usually near the surface in the open oceans' cold waters. There, many young, small fishes hide in my tentacles or under my bell. My poison doesn't hurt them. As I age, though, I seek shallow, sheltered bays.

Lion's Mane Jelly   15

# Bristle Worms

*smiiya-x* in Aleut • *Annelida Polychaeta (phylum/class)*

I am Dr. Wilma A. Worm, spokesworm for *all* marine bristle worms, the Polychaetes (*PAH-lee-keetz*). We have lived on Earth for more than five hundred million years . . . on beaches, in tide pools, and at all depths of all oceans. Marine worms are among the most common of all marine life with more than ten thousand species of us. There is no common worm, so I've been asked to tell you a bit about us.

Like most land worms, marine worms have long bodies made up of identical, tiny segments that look like little rings. Each segment, except for our head and tail, have the same sets of internal organs. New segments grow continually from the "growth zone" right in front of our tail. A tough, flexible skin-like covering protects our whole body. Setae (SEE-tee), or tiny hairs, stick out on this outer layer. Most of us also have "legs" called parapodia (pear-uh-PO-dee-uh) poking out on both sides of each of our segments. These paddle-like growths are covered with bunches of long or short, soft or hard, hair-like bristles. Swimming worms use their legs like paddles; walking worms use them to walk on the seafloor; and burrowing worms use their parapodia-like shovels to make their tunnels into the mud.

Marine worms are either free-moving, swimming or "walking" where they want, or they are tube-dwellers. Tube-dwellers' bodies secrete mucus that hardens when it touches seawater. Some worms attach their tubes permanently to rocks or other hard surfaces. Others hide their tubes in mud with only the opening above the seafloor.

Wherever we live, we all have an unsegmented head with eyes that only show us light and dark. (One species has huge eyes with all the parts of a human eye.) Other worms "sense" what is around using a pair of antennae on their heads. Some worms have a pharynx (FAIR-inks), a skinny tongue that turns inside out as it shoots out to grab prey. Depending on the food we eat, jaws may be huge or tiny.

Check out *these* worms! Zombie worms eat bones of decomposing animals, and fire worms have hot stingers to sting prey. Scale worms, the roly-polies of the sea, have strong, powerful jaws. And the Bobbit worm is a dangerous, meat-eating predator growing to be ten feet long! See! There certainly is no typical worm among the Polychaetes.

## My Facts

**SIZE**: Tiny enough to live between grains of sand to 10 feet long; most 0.3 inches to 4 inches. **COLOR**: Wide variety, from brightly colored and luminescent to gray, brown, green. **FOOD**: Carnivorous, herbivorous, omnivorous. This group of animals represents all types of eaters. **PREDATORS**: Over 100 organisms prey on worms, including crustaceans, fishes, shorebirds. **TIDAL ZONE**: From beach to ocean floor. **RANGE**: Worldwide. **LIFESPAN**: A few weeks to months to years.

## Did You See Me? Tell Your Story! _____

_____

_____

**DID YOU KNOW?** We live in *all* oceans and are very useful to scientists studying the health of marine ecosystems. If we aren't where we're usually found, why? What's wrong? Found too many of us? Why? Become a wormologist and study us!

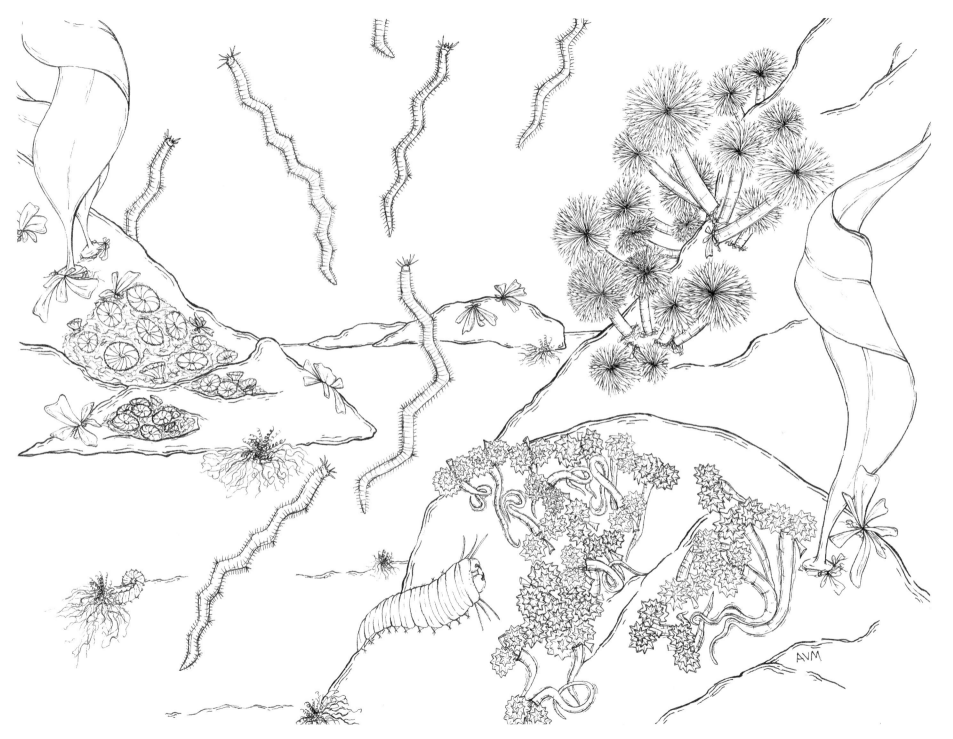

# Common Acorn Barnacle

*qauq* in Kodiak Alutiiq • *Balanus glandula*

Hey! Stand on your head, stick your feet in the air, then *eat* with your *feet!* Bet you can't. Well, I'm Barney Barnacle, and that's how *I* spend my entire grown-up life . . . head down, feet up. Plus, I'm *glued* to the same spot *forever*! I floated around after I hatched, and like most arthropods (AR-thruh-pahdz), I already had my exoskeleton. Wasn't long 'til I had to decide on a permanent spot. Did I want a dock, piling, ship's hull, or even a whale? Any hard surface would work.

I chose my spot on the top of this large rock in an intertidal pool, joining a group of about a thousand of my kind. Head down, my sticky antennae stuck to my rock and secreted brown, fast-setting glue. I was home. I started "building" my permanent, cone-shaped "house," or outer shell, directly on the rock's surface. I surrounded my body completely with six overlapping plates that make me look like a tiny volcano. My rounded base stuck to my rock with one of nature's most powerful glues. At the very top of my house is a small opening, and four more plates make a trapdoor to cover that opening.

When the tide comes in, my muscles pull that trapdoor open. Many pairs of slender, feathery, "leg" appendages called cirri (SEAR-ee) extend out into the water. My cirri make sweeping movements, combing through the seawater for microscopic plankton food particles. Once these feathery legs are loaded, other curly cirri move the food to my mouth. As the tide goes out, I stop feeding, my cirri comes back in, and strong muscles pull the plates closed. This keeps my soft body from getting too warm and drying out when I'm above water.

Closing my top plates protects me from predators, too. Sea snails try to drill through my trapdoor and exoskeleton to eat my soft body. Ribbed limpets want to bulldoze me off my rock. Ochre sea stars use their tube feet to break my shell apart. Worms try to get inside my shell, and seabirds try to pick me off rocks. There is danger everywhere, so I need all the protection I can get!

Ships need protection from *us*. As many as ten tons of us will attach to a ship's hull in less than two years. We add weight, plus make bumps on the ship's smooth hull adding drag, or resistance, when the ship is moving. This may increase fuel usage as much as forty percent! Little me, big problem!

## My Facts

**SIZE**: Diameter: 0.5 - 0.9 inches. Height: usually same as diameter. (If we're crowded, we grow taller with a smaller diameter.) **COLOR**: Outer shell: white or grayish white; may have longitudinal ribs. **FOOD**: Omnivorous. Zooplankton, detritus. **PREDATORS**: See story. **TIDAL ZONE**: Most in mid and upper intertidal (some live above, some live below). **RANGE**: Aleutian Islands, Alaska, to Baja California, Mexico. **LIFESPAN**: +/- 10 years.

## Did You See Me? Tell Your Story! _____

_____

_____

**DID YOU KNOW?** Acorn barnacles are the most common of *all* barnacles, and there are more of us than any other species! I'm closely related to shrimps and crabs. In fact, when I was born, you'd have thought I was a tiny little shrimp!

Common Acorn Barnacle

# Seaweed Isopod

*chagliiqnaang* in Aleut • *Pentidotea wosnesenkii*

Gotta talk fast and stay hidden! Izzy Isopod (EYE-so-pod) here, a small, bug-like, marine crustacean (kruh-STAY-shun). I'm the ocean's roly-poly pill bug, the "sow bug of the sea," even though I never roll up like land isopods. I'm one of the largest isopods in my zone, an easy target for birds, fishes, and marine mammals. So I hide under rocks, in mussel beds, empty sea snail shells, on floating docks, and especially among seagrasses and seaweeds. I'm slow-moving and slow-swimming, so I *must* stay out of sight!

Right now, I'm hiding in this dark, olive-green kelp, or seaweed. I'm the same color as this patch of weeds, so I'm camouflaged and safe. Sometimes my kind will be brighter green, dark gray, black, brown, or even pinkish. Why? Well, like all crustaceans, we molt, shedding our exoskeletons as we outgrow them. As new exoskeletons form, we change colors to be more like the color where we're hiding and feeding. If I'm on the seafloor, which is more brown than green, I become more brown. If the kelp is a brighter green, that's the color I become. My little body operates just like that reptile, the chameleon. I camouflage to match where I am.

My colorful body is little. It's rectangular shaped, flattened with a little arching in the middle, and definitely sturdy. I have seven tiny segments, each covered with an overlapping plate of my exoskeleton. These plates protect me from predators and make it easy for me to move. On each body segment I have an identical pair of legs, one leg on each side. (My name "isopod" comes from *iso* meaning "equal" or "same," and *poda* meaning "foot.") All my legs are for crawling. The last five pairs help me swim, too. No pinching claws on the end of my legs, like crabs have. Mine just curve under a little and end at a point. See why I cling so easily to rocks, docks, seaweeds, your finger? Don't worry. It won't hurt if I *do* grab you.

I *am* small, so look carefully for me! If you see me coming *toward* you, look for two, kidney-shaped eyes on the sides of my head. On top of my head are two pairs of antennae for finding food. I smell for prey with the short pair, and the longer pair feels around for a victim. If you see me going *away* from you, you'll see my tail with its outward curve and small, blunt tooth right in the middle. And right now, I absolutely *must* get going and stay *completely* out of sight!

## My Facts

**SIZE**: Length: up to 1.6 inches. Width: up to 0.5 inches. **COLOR**: See story. **FOOD**: Mainly herbivorous. Macro-algae, seagrasses, sea snail's egg capsules. **PREDATORS**: Small fishes, seabirds, marine mammals. **TIDAL ZONE**: Intertidal to 54 feet deep. **RANGE**: Central California to Alaska to Russia. **LIFESPAN**: 3 - 4 years.

## Did You See Me? Tell Your Story!

_____

_____

_____

**DID YOU KNOW?** I grew from an egg carried in a pouch of fluid on the underside of my mom's belly. The pouch protected me and kept me from drying out if Mom was above sea level. When I hatched, tiny-me looked just like I do now.

Seaweed Isopod   21

# Common Beach Hopper/Sand Flea
*kunt'gwaang* in Haida • *Traskorchestia traskiana*

*Hey*! Who's messin' with my hideout? Hooper A. Hopper here, and you'd better *not* poke around in my dark, damp, rotting, high-tide line of grasses and weeds. You move my home, food supply, *and* place where I hide from dangerous beasts? You'll see a massive explosion of hoppers hopping and jumping like you've *never* seen! *Completely* scatter my washed-up line of weeds? I'll burrow about an inch into the sand where it's dark and moist. You'll *never* see me unless you sneak back when the tide's out and it's dark. That's when *thousands* of us will be out on the beach whoopin' it up, crazy hopping, and scavenging for food.

Yup, there are thousands of species of us, many living in huge groups on sandy or pebbly beaches in protected areas. Others of us live in the ocean, from shallow water to as deep as it goes. We are small crustaceans with segmented, shrimp-like bodies, taller than we are wide, narrow and arched. We're amphipods (AM-fuh-podz), which means "different feet." Different species of us have different types of legs for varied uses. Even my *own* legs aren't all the same. My front pairs are the longest and are made for grabbing food. The next three pairs are shorter, made for walking and swimming. (I'm a lousy swimmer, but I don't talk about that.) My hind legs? *Those* are my hopping legs, and I *definitely* talk about them. My legs curve backward, up, and over my back. When I flex them, then straighten them . . . *SNAP*! It's like having springs that shoot me as far as a *foot* into the air! The big hopper here, ready to leap in a single bound!

Unfortunately, I've been known to *overshoot* my mark and hop way above or below my high-tide line of weeds. No prob! I have special powers that take me right back to my usual spot. I use my complex eyes to check the angle of the sun, like using a compass. Then my GPS-like "map sense" kicks in to check beach slope, sand moisture, and grain size. I go right *back* to my own exact spot!

Take note that I'm not just a useless, jumpy guy. Amphipods are hugely important to the marine ecosystem. We're decomposers, eating dead stuff in the sea and on shore to get rid of it. Plus, my kind is a major food source for large and small creatures in the ocean. Without us, even those huge gray whales might starve. I'm *small* but a *big* deal. Got it?

## My Facts

**SIZE**: Length: up 0.75 inches. **COLOR**: Body: creamy brown, brown, gray. Antennae: transparent. **FOOD**: Mainly herbivorous. Tiny pieces of algae, seaweeds, eel grasses; some decaying plant/animal parts. **PREDATORS**: Fishes, whales, shorebirds, foxes, bears. **TIDAL ZONE**: Splash zone to high intertidal. **RANGE**: Aleutian Islands, Alaska, to Baja California, Mexico. **LIFESPAN**: 1 year.

## Did You See Me? Tell Your Story! _____
_____
_____

**DID YOU KNOW?** Sometimes I'm called a "scud" or "side-swimmer," which works for me. Why I'm called a sand flea, though, I have no clue! I'm *NOT* a close relative of those insects, nor do I *ever* bite and suck blood from anyone. That's *gross*!

*Common Beach Hopper/Sand Flea* 23

# Spot Shrimp

*s'eex'at* in Tlingit • *Pandalus platyceros*

It's quiz time! What's the difference between a shrimp and a prawn? Give up? Well, I'm Sharla Spot Shrimp, and I *am* closely related to prawns. We're both crustaceans, look like twins only I'm smaller, and our curvy bodies have a few, tiny differences. *You* won't be able to tell us apart if you eat us for dinner. But you *must* know, *I'm* the largest shrimp on the entire west coast, usually the one in fish markets, freezers, and restaurants. In those places, they've already cut off my head and stripped my exoskeleton. *You* only eat my long, slender, muscular abdominal part. The whole of me, in my ocean world? I look *so* different!

My drawing shows all of me. On my head, I have two, black eyes and two, very long antennae that look like overgrown whiskers. See my rostrum (RAH-strum), the long, hornlike extension between my eyes? This hairy, pointy part is the front of my carapace (CARE-uh-puhs), the exoskeleton covering my head and thorax (THOR-ax), or chest. You must see my most important feature! White spots, right behind my head and in front of my tail. *Those* spots gave me my name!

When you only see the "ready to eat" me, you'll miss my slender, fragile legs that attach to my exoskeleton. The first three pairs help me feed; the very first ones even have claws! Next, I have two pairs of walking legs. Spot this! The left leg of my second pair is longer than the right one. This long left leg is totally unique to spot shrimp. The rest of my legs, all the way back to my tail, are my swimmerets (SWIM-uh-rets). Swimming is how I get around, so I use these legs all the time. I paddle to move forward, but if a predator comes near, I retreat super fast by repeatedly flicking my tail.

October is when I make a big retreat to deeper water. By then, I've molted and grown a new shell. This new exoskeleton has a special pouch on the abdomen where my newly laid eggs have attached. All winter these eggs grow, held firmly against me by tiny, hair-like setae (SEE-tee). Finally, in April, my eggs hatch into free-floating, tiny larvae. Slowly, these larvae grow into juvenile shrimp and drop to the ocean floor. They continue to grow, becoming adult, male shrimp. But as they grow more . . . surprise! Most of them change into female shrimp. Isn't that absolutely amazing?

## My Facts

**SIZE**: Length: up to 12 inches (female usually larger). **COLOR**: Body: reddish brown or tan. Carapace: white horizontal bars. Swimmerets/antennae: banded dark red, lighter red, or white. **FOOD**: Mainly carnivorous. Worms, diatoms, dead/decaying organisms, other shrimps, small mollusks, sponges, algae. **PREDATORS**: Humans, large fishes, octopuses, marine mammals, seabirds. **TIDAL ZONE**: Intertidal to 1,600+ feet deep; most near 360 feet. **RANGE**: S. California to Aleutian Islands, Alaska; Sea of Japan to Korea Strait. **LIFESPAN**: 7 - 11 years.

## Did You See Me? Tell Your Story!

**DID YOU KNOW?** I'm an important food source for many animals, even humans! Fishermen all along the coast of North America earn about fifty billion dollars a year catching me. But pollution and warming waters are killing us. Please help!

# Hairy Hermit Crab

*skats'gw* in Haida • *Pagurus hirsutiusculus*

Hey! Harry's the name, Hairy Hermit Crab. I'm one *very* hairy Harry with hundreds of small, stiff hairs all over my legs and carapace. I just moved my hairy body into this empty Sitka periwinkle's shell, and I love it! So, you know how barnacles and most other crabs and crustaceans make their own outer shells? Well, I don't. My hairy little body, with its soft abdomen, is unprotected. Solution? I just move into someone else's abandoned home, a ready-made fort to use against attackers. Sea snails' shells are super! And since I just molted out of my old, tight exoskeleton and grew a new one, I needed this larger shell.

I didn't leave my old shell until I found this new, larger one. It's sturdy, and the right weight and fit. I checked the shell's strength with one of my chelae (KEE-lee), my front legs with claws and pincers on the ends. I poked my leg inside, and the setae (SEE-tee), or bristly hairs on the claw, touched around to see if the shell was hard and firm. Then the size. It had to be small and easy to carry. I have to move fast when danger's near, and a heavy shell slows me down. My speed often saves me, so I'll even drop my shell if it means staying alive.

With a small shell, much of my body sticks out. Don't worry. The shell won't fall off. My abdomen is soft, and it winds perfectly into the curling chamber inside the shell. Then, my abdominal muscles press hard against that shell's interior walls. When the hooklike part on my left, rear foot grips the inside column, I'm in there tight! If I have to hide from danger, I scrunch as much of my body inside as I can. Most fits, and I use my bigger claw on the chela on my right side to block the shell's opening. A built-in trapdoor on my portable cave.

I'm like a hermit hiding out in my cave. When I'm *not* hiding, though, I'm like all crabs. I wear my shell as I crawl on the bottom of sandy or rocky tide pools, or in shallow, protected waters along shorelines. I scuttle along using my two claws and front-most pairs of walking legs. I'm usually prowling for rotten, decomposing seaweed and dead animals to eat. My long, segmented antennae feel around rocky areas, and when I find something, my larger right claw grabs and holds the food. Then, my smaller left claw breaks off pieces and moves them to my mouth. The more rotten, the tastier. And that, my friend, is livin'!

## My Facts

**SIZE**: Body: up to 2.8 inches (male usually larger). Carapace: up to 0.75 inches. **COLOR**: Body: overall olive green, brown, or black. Legs: see Did You Know? Antennae: green or grayish brown with yellow or white bands or spots. **FOOD**: Omnivorous. Mainly detritus, seaweeds, live snail hatchlings. **PREDATORS**: Fishes, especially tide pool sculpins; parasitic barnacles. **TIDAL ZONE**: Mid to upper intertidal to 365 feet deep. **RANGE**: Pribilof Islands, Alaska, to S. California; Siberia to Japan. **LIFESPAN**: Unknown.

## Did You See Me? Tell Your Story! _____

_____

_____

_____

**DID YOU KNOW?** Look for my hairy body, tiny shell, and banded antennae. Then find a blue dot at the top of each leg, a white band on the outer segment of my walking legs, and a blue patch on the tips of my legs. Yup, that would be me!

Hairy Hermit Crab

# Dungeness Crab

*tiq'achihi* in Dena'ina Qenaga • *Metacarcinus magister*

Hellloooo! It's me, Courtney Crab, a Dungeness crab. You *know*, I'm the largest, most abundant crab on the entire Pacific coast. Oh, alright. Crayton King Crab is much bigger, but kings only live far west of Alaska and northern Canada. *I'm* the prize everywhere else. Commercial fishermen earn millions of dollars each year harvesting us. Thousands of regular people trap us in crab pots or ring nets. My sweet, moist, tender meat is a "must" for you crab-lovers. But, alas, do you *really* know anything about me, wondrous crustacean of the deep that I am?

I'm a crawler who loves the soft sand and mud of the ocean's floor as well as smooth eelgrass beds. I *never* swim! I use my four pairs of short, walking legs to move sideways. Four legs on one side pull while the four on the other side push. I even *run* sideways when I need to get away from some horrible enemy and don't have time to burrow backward into the sand. My two strong, grasping claws are out in front. The white-tipped pincers on the ends have jaggedy, toothy edges, simply *maaavelous* for fighting off predators. I use these claws mainly to poke in mud and sand for food, however. And oh, when I find a clam or worm? My pincers just *tear* that tasty creature into pieces. Then my small feeding appendages pass the morsel down to my mouth. That's under my head at the front of my flattened body.

I'm *sure* you noticed the most beautiful, reddish, flattened, oval-shaped carapace covering my body. This shell is so *rudely* ripped off when you buy me whole at the fish market! If you look around the edges of that shell you'll see my small, ridge-like "teeth." The last, largest tooth points directly outward. Only we "*Dungies*" have that unique tooth!

I *am* unique! I molt my old, outgrown exoskeleton each year as my new one begins to form. I move into shallower water and slide out of my old carapace through a slit in the rear. Once out, I drink enough seawater to stretch my new, pliable exoskeleton into a larger size. I need room to grow! When it's completely formed, I bury myself in sand for several days while my new covering hardens. Soon, I'm ready to go, side-stepping again. If you're on the beach when we're molting, you might find outgrown exoskeletons washed up on shore. No worry. We're fine.

## My Facts

**SIZE**: Length: up to 6.5 inches (male usually larger). Width across back of carapace: up to 10 inches, usually under 7.9 inches. (In Alaska, only males at least 6.5 inches across may be kept.) **COLOR**: Carapace/top of legs: light reddish brown or grayish with purplish wash; lighter streaks and spots. Underside: whitish yellow to light orange. **FOOD**: Carnivorous. Adults: variety of small fishes; small invertebrates like clams, mussels, worms, shrimps; dead marine organisms. Juveniles: fishes, shrimps, mollusks, other crustaceans. **PREDATORS**: Humans, sea otters, octopuses, sculpins, herring, halibut, other crabs. **TIDAL ZONE**: Low tide line to 750 feet deep; most abundant less than 300 feet deep. **RANGE**: Aleutian Islands, Alaska, to Baja California, Mexico. **LIFESPAN**: 8 - 13 years.

## Did You See Me? Tell Your Story! _____

_____

_____

**DID YOU KNOW?** If I lose a leg, it regenerates, or regrows, during molting. It's smaller after the first molt but larger with each yearly molt until it's normal size. It would be so wonderful if you humans could regrow your body parts!

# Pygmy Rock Crab

*puyyugiaq* in Inupiaq • *Glebocarcinus oregonesis*

Yawwnn! Crabby Cray Crab here, and I'm just heading off to bed. I've been out all night eating barnacles and other small invertebrates, and I'm *tired*. It's morning, so I'm ready to tuck myself way back inside this empty, giant barnacle shell. Once I'm all in, I'll plug the opening with my carapace, or flat, back shell. I'm going to sleep the day away. Sometimes, I sleep in small holes in rocks, under rocks, or in small crevices between rocks. This rock crab digs rocky places!

My rocky places are along the west coast of the Pacific Ocean, including Oregon. See my species name, *oregonesis*? That led to other names for me, like Oregon Rock Crab and Oregon Cancer Crab. Cancer? No, no. I don't *have* cancer. My genus name *Glebocarcinus* ends with *carcinus*, meaning "hard shell." I'm one of the group of eight cancer crabs who have hard shells with spiky "teeth" around the outside edge. Like the other seven, I'm a decapod with ten legs, and I never swim. Most of my legs are made for walking, while my front legs have short claws for eating. Do you see my short, pointy rostrum, the front extension of my carapace? Cancer crabs *all* have a rostrum sticking out between wide-set, popping eyes that stand up on short stalks.

That's a good description of a cancer crab. But there's a *big* problem! Cancer crabs are *big* crabs! Huge Courtney Dungeness? Without question, she's a cancer. Now look at me! I'm *tiny*, a pygmy (PIG-me) crab! My carapace measures no more than two inches across at its *widest* point! And, before you ask, *NO*! I'm not a miniature Courtney. I won't grow up to be a big Dungeness crab.

I'm my own, unique, tiny crab! Unlike the other cancer crabs, I have an almost perfectly round carapace. That shell is covered with small bumps that look like warts. Parts of my claws have warts, too, not spiny ridges like the other cancers have. On the ends of my claws are short, black-tipped pincers that are so powerful they pry open rock-hard barnacle and gastropod (GAS-truh-pod) shells. And look closely. I'm *really* hairy! I grow dark, grayish-green setae (SEE-tee) on the outer surface of my mighty pincer-like claws, all over my short, walking legs, and along the edges of my carapace.

That's enough! I'm tired and need my beauty sleep. Good night.

## My Facts

**SIZE**: Length: up to 1.3 inches (female larger). Width: up to 2 inches. **COLOR**: Body: overall dullish red, red brown, reddish orange; sometimes mottled, striped, or lighter. Underside: lighter whitish; Tip of pincers: black. **FOOD**: Carnivorous. Mainly small barnacles, mussels, snails, worms, other small crustaceans. **PREDATORS**: Pacific cod, river otters, red rock crabs. **TIDAL ZONE**: Low intertidal to 1,400 feet deep. **RANGE**: Pribilof Islands, Alaska, to S. California. **LIFESPAN**: Unknown.

## Did You See Me? Tell Your Story!

_____

_____

_____

**DID YOU KNOW?** If I'm in danger, I tuck my legs up tight and roll away like a round pebble. No other cancer crab does that! Look for tiny me at low tide on rocky beaches, but I also hide in mussel beds and on pilings. I'm hard to spot!

Pygmy Rock Crab   31

# Graceful Decorator Crab

*(Can you find a Native name for me?)* • *Oregonia gracilis*

As you will learn, my name tells you *exactly who* and *what* I am. I'm Graciella. Graceful. Decorator. Crab. I'm a spider crab, made *quite* obvious by my two pairs of very long, slender, delicate, spider-like walking legs at the hind end of my body. I *am* graceful when I walk smoothly on these long legs. My front three pairs of legs are a bit more plump, perfect for finding food and moving it to my mouth. My very front pair, with claws and pincers, are my chelipeds (KEE-luh-pedz). I use them to find food, and also to defend myself when *that* becomes necessary. But most importantly, these . . . are my decorating legs.

I'm a fashionista, one of nature's most fashion-conscious animals with a *highly*-developed decorating habit. With *great* attention to detail, I decorate myself to blend in with my surroundings. I camouflage my body *carefully*, becoming invisible to nasty predators who could easily snatch and eat my slow-moving body. My decorating pincers nip off pieces of seaweeds, sponges, and algae, or grab small, polyp-sized invertebrates, like baby anemones. I make use of almost *everything* around me. I attach the pieces to fine, Velcro-like hooked hairs on my back and legs. The hooks hold like glue! What's *terribly* convenient is when I move to a new locale, I'm able to change my outfit. I pull the old scraps off and select all new items to hide me in my new environment.

My unique decorating keeps predators from seeing me, but some parts of my outfit are actually *dangerous*. I find pieces of seaweeds that are poisonous to some enemies, and *I* know which tiny anemones sting. I attach algae that makes me taste terrible, *certainly* not like the tasty crab I normally am. *Nobody* wants to eat me then!

*Whatever* my flavor, I *am* a crab! My small, rounded triangular carapace is wider in back than in front. My rostrum pokes out in front then splits into two long, parallel spines. Two long horns stick up behind each of my stalked eyes, so it looks like I have teeth on top of my head. And just like *other* crabs and crustaceans, I molt out of my exoskeleton when it becomes too tight. What happens to all my fancy decorations when I leave my shell? I simply remove my old accessories and attach them to my *new* shell when it hardens. Incidentally. How do you like my *latest* "camo" outfit?

## My Facts

**SIZE**: Length: up to 3 inches including rostrum. Width: half of length. **COLOR**: Body: brown, tan, or gray under decorations. Feet: white with pink or orange tips. **FOOD**: Omnivorous. Floating kelp, algae, sea urchins, sea anemones, other echinoderms, carrion. **PREDATORS**: Pacific halibut, octopuses, sea otters. **TIDAL ZONE**: Intertidal to 1,430 feet deep. **RANGE**: Bering Sea to Monterey, California; Japan. **LIFESPAN**: Unknown.

## Did You See Me? Tell Your Story! _____

_____

_____

**DID YOU KNOW?** I live alone on rocks, sand, and gravel in tide pools along the shore. I also hide on the deep seafloor. At night, I look for food, but during the day, I'm still and quiet. That's when most predators try to find me. Good luck!

Graceful Decorator Crab

# Mottled/True Sea Star

*sk'aam* in Haida • *Evasterias troschelii*

I'll bet you know who *I* am! Just look at my drawing and you'll know that I'm a symbol of the sea *everyone* recognizes. Any of your kind who go to rocky beaches or tide pools *always* want to see Sebastian Sea Star, the Mottled or True Sea Star. I'm actually just one of more than two *thousand* species of sea stars. You have a lot of stars to see!

Notice I'm saying "sea star" not "star*fish*?" You know why? We're *not* fish, just like jellies aren't. Sea stars don't have a backbone, gills, scales, or fins. Scratch heart, blood, and lungs off the list. Fish have *all* those, *and* they have only two sides. Yes, I have an aboral (AH-buh-ruhl), or upper side, and an oral, or underside. But I also have at least five rays sticking out from a round disc right in the center of my body. Some stars even have twenty rays, or *more*! Have you ever seen a fish with rays and a central disc? *Ha!*

Me, I'm a pretty regular sea star with my five long, slim rays. They sort of curl up at the tips and are widest a little bit away from my small center disc. Check out my skin. Looks lumpy? It is! If you touch me, you'll feel spiny bumps. I'm actually like most sea stars and other critters in the echinoderm (eh-KAI-nuh-durm) phylum with rough, pokey, "hedgehog skin." The spines on my aboral side are in clumps and bunches, and underneath, I have row after row of spines. From the tips of my rays all the way to my mouth in the center of my disc, spines. They're good protection from *dangerous* sea stars and other predators. Even better are my tiny, hard, pinching jaws around the spines. These jaws look like bird beaks, and man, do they bite hard!

Speaking of hard, remember how crabs, shrimps, and other crustaceans have a hard *exo*skeleton on the outside of their bodies? Well, under our spiny skin, echinoderms have an *endo*skeleton, or inside bony structure. It's made from lots of tiny, bone-hard plates held together by really strong, bendy tissue. This skeleton holds my shape but lets me move easily.

Don't expect to see me move, though. I'm slow. I take my time as I look for food in eelgrass beds or on rocks and pebbles in calm, intertidal bays and tide pools. I'm slow, but surprisingly dangerous as I slyly and sneakily move up to my victim. Gentle I'm not when I pull off my surprise attack. They *never* knew what hit 'em!

## My Facts

**SIZE**: Diameter including rays: up to 12 inches across when intertidal; up to 24 inches across when deeper in ocean.
**COLOR**: Aboral side: gray, greenish gray, bluish gray, gray brown, reddish brown, brown, or pale purple; may be plain or mottled. Oral side: pale brown. Rays: some have rows of white spines; color of outer edges often contrasts with rest of ray. **FOOD**: Carnivorous. Bivalve and gastropod mollusks, including chitons, limpets; some barnacles, tunicates.
**PREDATORS**: Humans, gulls, morning sun stars, sunflower stars.
**TIDAL ZONE**: Low intertidal to 246 feet deep.
**RANGE**: Pribilof Islands, Alaska, to Monterey Bay, California; uncommon south of Puget Sound, Washington; largest intertidal star in Juneau area.
**LIFESPAN**: Up to 35 years.

### Did You See Me? Tell Your Story! _____
_____
_____

**DID YOU KNOW?** I'm usually alone, but two tiny animals live on me for protection *and* to eat the scraps of food I drop. Baby Alaska king crabs like to hide between my rays. Scale worms stay in my underside grooves and keep them clean.

# Sunflower Star and Morning Sun Star

*kamadyaali-x and igadaga-x* in Aleut • *Pycnopodia helianthoides and Solaster dawsoni*

*What* were you thinking? You put *me*, Sunny Sunflower Star, in front of the Death Star of the Ocean! Don't you know that Monroe Morning Sun Star is a horrible, vicious predator, a *cannibal*, who eats *everyone*? His diet is almost completely sea stars, including his *own kind*, and . . . *ME*! Yes, I'm a serious predator, too, fighting with other sunflower stars I meet. Ok, ok. I *do* eat *some* stars, but I'm not *nearly* as bad as Monroe! He's the most dangerous and feared of *all* sea stars!

Maybe he hasn't seen me yet. Monroe moves incredibly slowly, and I'm the *fastest*, largest, and heaviest sea star in the North Pacific. If I need to, I travel at least six feet per minute on the thousands of tiny, tube feet covering the oral side of my tapering rays. I have anywhere from sixteen to twenty-four rays, almost *double* Monroe's eleven or twelve. I actually dropped a few rays recently getting away from enemies, like *Monroe*. I still have more rays than *any* other star, and my missing ones will completely regrow in a few weeks. It's better that he eats *one* ray than *all of me*!

I have other built-in ways to protect myself. My broad, bulky, limp body is covered with little, pinching jaws that nip at enemies as well as clear debris off me. Spiny projections on my aboral side and all along the sides of my rays poke predators who come too close. What's *really* important is my inside skeleton with its loose plates, giving me amazing flexibility when I move. Being flexible is why I'm so speedy when I need to escape as well as when I chase prey.

Once I catch my victim, I move my entire body on top of it. The loose skeletal plates around my mouth disconnect, so it opens wide enough to swallow food whole. I also eat like other sea stars using my *two* stomachs. Say I grab a tasty clam. I center my mouth over it and surround its two-sided shell with my rays. My suction-cup tube feet move in, hold on tightly, and barely pull those shells apart. In a flash, I spit my front stomach out through my mouth into that clam. Chemicals from my stomach turn the clam's body into runny goo. I suck my stomach and the gooey clam back inside my body for my *other* stomach to digest that runny clam.

Oh *NO*! Monroe's *looking* at me! I have gotta get out of here before I, too, become *goo*!

## My Facts

**SUNFLOWER STAR**: **SIZE**: Diameter including rays: up to 39 inches. Weight: up to 11 pounds. **COLOR**: Aboral: varies; often reddish orange to yellow; also, red, brown, purple. Oral: bright yellow orange. **FOOD**: Carnivorous. Sea urchins, mussels, clams, crabs, barnacles, chitons, sea cucumbers, snails, fishes. **PREDATORS**: Humans, morning sun stars, Alaska king crabs, sea otters, gulls, bottom fishes. **TIDAL ZONE**: Low intertidal to 1,435 feet deep. **RANGE**: Unalaska Island, Alaska, to Baja California, Mexico. **LIFESPAN**: 3 - 5 years. **MORNING SUN STAR**: **SIZE:** Diameter including rays: up to 16 inches. **COLOR:** Aboral: red, orange, gray, yellow, brown with paler patches. **FOOD:** Carnivorous. Sea stars, sea cucumbers, diamondback nudibranchs, anything else it wants. **PREDATORS:** Humans, other morning sun stars, Alaska king crabs, sea otters, gulls, bottom fishes. **TIDAL ZONE:** Low intertidal to 1,380 feet deep. **RANGE**: Alaska to S. California; Japan, China, Siberia. **LIFESPAN**: Up to 20 years.

## Did You See Me? Tell Your Story!

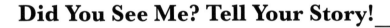

**DID YOU KNOW?** We both live on rocky, gravelly, or sandy sea floors. If our victims "sense" that we're around, they try to run or swim away. If we grab them, they twist violently or poke us with long, sharp spines. Can you guess who wins?

Sunflower Star and Morning Sun Star   37

# Blood Star

*s'ax* in Tlingit • *Henricia leviuscula*

Oh! Hello! I'm down here on this big rock relaxing in my favorite tide pool. I'm Starlynn Star, a brilliant blood star you *must* color the brightest red or reddish orange you have. I make *such* a splendid color splash in protected pools, or anywhere I find calm, peaceful seawater. I'm a quiet soul, living alone on and under rocks, or in underwater crevices or caves. I'm quite graceful with my five long, slender, tapering rays turned up slightly at their tips. If you were to touch my rays or small, flat central disc, you'd feel a smooth but gritty surface, like very fine sandpaper. That sandy feel is the tiny, blunt ossicles (AH-sic-kuhls) covering my aboral, or top side. To see those tiny, hard spines you'd have to use a microscope!

It's also hard to see the tiny tube feet on the oral side of my slim rays. Each of my rays has a narrow, deep furrow, or groove, right down the middle from its tip to my mouth. All along the grooves are rows and rows of tiny feet. I have *hundreds* of feet, and stars with longer rays have *thousands* of feet! Every single little foot is a suction tube, and every one helps sea stars and all echinoderms (eh-KI-nuh-durm) move, eat, and stay alive!

Echinoderms are the *only* animals who have a "water vascular system." What's that? It's a bit like your heart and blood. I pump seawater into my body through a tiny, light-colored opening near the center of my upper side. The water goes into a pipeline of tubes, and my muscles help move that water through my disc and into all my rays. It's like blood being pumped throughout your body. The pumping causes pressure that moves my rays, helping *me* move. The seawater also goes in and out of my tube feet, making them grab a surface tightly or let it go. That's how I hold onto or let go of rocks. More importantly, it's how I hold onto prey!

Most sea stars eat their prey by turning one stomach inside out to trap, then liquify, or make goo out of their victim. I do that when I find larger pieces of food. More often I use an easier, cleaner style. My body is covered with sticky mucus and tiny, little cilia (SIL-ee-ah), or hairs. When microscopic bits of sponge, plankton, or bacterium float close to me, they stick to my mucus. Once caught, my cilia gently move the food to my mouth. Voila! A quick, tasty snack.

## My Facts

**SIZE**: Diameter including rays: rarely larger than 5 inches. Central disc: less than 20% of diameter. **COLOR**: Aboral side: usually bright red or reddish orange; sometimes tan or purplish; some mottled gray or gray-violet "ray-pits." Oral side: lighter yellowish. Rays: may have darker bands; not mottled. **FOOD**: Carnivorous. Sponges, bacteria, bryozoans, other tiny particles. **PREDATORS**: Humans, birds, other stars. **TIDAL ZONE**: Low intertidal to about 1,300 feet deep. **RANGE**: Aleutian Islands, Alaska, to Baja California, Mexico; Japan. **LIFESPAN**: Up to 35 years.

## Did You See Me? Tell Your Story!

_____

_____

_____

**DID YOU KNOW?** When I'm turned over on my back, I may die! It takes at least *fifteen minutes* to turn right-side-up. I move two rays toward each other, lean on two others, then flip over using my last ray. It's called the "sea star flip." Tricky!

38 Blood Star

# Ochre Star

*agyaruaq* in Kenai Alutiiq • *Pisaster ochraceus*

I am one important sea star! Oleg Ochre's the name, a Keystone species. *That* means what I do affects other marine animals and the health of our intertidal environment. You know that California blue mussel, the pesky, blue-black bivalve, or two-shelled, marine animal? Often, his kind *completely* takes over large areas of our intertidal homes, eating *every* bit of algae and seaweeds. Those maddening mussels leave no room or food for any other marine animals to live there. Mussels completely *destroy* the balance of our environment!

*Well*, it just so happens, blue mussels are my absolute *favorite* food. I eat up to eighty of them each year. So, when enough ochres eat enough mussels, their population is controlled, making our intertidal world *much* healthier. When you see mussels clustered up high on rocks or pilings, you can *bet* I'm below them. I stay cool in the water while letting them know I'm there, waiting patiently to make a meal of them. Yum!

Mussels *know* to watch for my fairly large, thick, heavy body with its slightly raised central disk. I'm the typical sea star with five, up to seven, plump rays that thin out as they taper toward the ends. Covering my aboral side are hundreds of tiny, short, white ossicles, or spines. Set in close rows, my spines make unique, net-like patterns all over my colorful body.

Ochre, or orangish, is the color many of us wear. We're also yellow to reddish to brown, and my favorite, purple. That color gave us another name . . . Purple Stars. Someone else named us Common Stars, probably because you see so many of us clustered on rocky shores where waves crash. You'll see large groups of us, our suction feet clinging tightly to boulders or rocky ledges. Another of my favorite places is in cold, shallow tide pools. But the *best* place of all? Right in the middle of mussel beds.

Yes, I'm the predator there, but I'm also preyed upon. Often I'm injured or even *lose* a ray. Recently, I dropped one Monroe Morning Sun Star had grabbed. But like all sea stars, I have *super* powers and regenerate, or regrow, a missing or injured ray. I even regenerate my entire body if there's a ray or part of my disk alive. Stars have all their critical internal parts in each ray and central disk. Complete regrowth may take a year, but that's how we live so long!

## My Facts

**SIZE**: Diameter including rays: up to 13 inches. **COLOR**: Aboral side: see story. Oral side: lighter. **FOOD**: Carnivorous. Mainly mussels; also, barnacles, snails, limpets, chitons. **PREDATORS**: Humans, glaucous-winged gulls, sea otters. **TIDAL ZONE**: Mid intertidal to about 290 feet deep. **RANGE**: Prince William Sound, Alaska, to Baja California, Mexico. **LIFESPAN**: 20+ years.

### Did You See Me? Tell Your Story!

_____

_____

_____

**DID YOU KNOW?** In 2013, *millions* of ochres died from Sea Star Wasting Disease. Our rays fell off, and we slowly disintegrated into white goo. Warmer water temperatures from climate change are a major factor. Will you help save stars?

# Leather Star

*krüsta-x* in Aleut • *Dermasterias imbricata*

Do you smell something a little stinky? Shhhh! Don't tell anyone! You see, it's me, Santino Sea Star, a gorgeous, bluish-gray leather star with a slight body-odor issue. I smell like garlic with a hint of burnt gunpowder. My fragrance doesn't bother me, but it signals others that I'm here. In fact, there's a certain, tasty sea anemone who panics when I come near or barely touch it. (Find out who *that* is!) It freaks out, quickly lets go where it's attached, and thrashes madly as it swims away. I don't even *try* to catch that runaway anemone.

Sea stars like me *never* swim, and I walk *amazingly* slowly. My thick, broad body, with its five, short, plump rays, isn't built for speed. My rays widen from their tips to where they attach to my thick central disc. There's webbing where my rays attach, like the skin between the toes of a duck. This adds even more to my padded appearance.

Covering my padding is my smooth, shiny smelly skin. It's a bit slippery from the slimy mucus oozing out. Touch me, and I feel like . . . wet leather! My aboral surface doesn't have spines or pinching jaws like other five-rayed stars, but I *do* have lumps. These are my skin gills, small, rounded papulae (PAP-yuh-lee). All sea stars have some type of skin gills to breathe oxygen out of seawater when the tide comes in to cover us. We need that oxygen to live!

Sea stars all have red or black eyespots, too, in the skin at the tip of each ray. My eyespots "see" enough light and dark to recognize shapes, but I don't see color or details. Shapes are good, though, when I'm searching for food. When I do sneak up on an anemone, my favorite, I usually swallow the whole critter in one, big gulp. Oh, I *totally* can do the "spit-out-my-stomach-dissolve-the-victim" approach, but *I* prefer to digest my food at a more relaxed pace.

I'm sure *not* relaxed if I hunt for food, and deadly Monroe Star is lurking around. He tracks my garlicky smell all too easily. I usually stay in protected water, hiding under rocks or holding onto pilings and sea walls. And like all stars, I have rows of tube feet on the oral side of my rays, with spines along the grooves where my feet line up. But they don't cling to rocks very tightly, and being a slow-mover, I'm an easy target. If I'm not careful, I become prey, *not* predator! *Help*!

## My Facts

**SIZE**: Diameter including rays: rarely larger than 12 inches. Thickness: up to 4 inches. **COLOR**: Aboral side: blue gray covered with net-like pattern in reddish brown, orange. Oral side: yellowish. **FOOD**: Carnivorous. Mainly sea anemones; also, sea cucumbers, sea urchins, sea pens, chitons, sponges, bryozoans, fish eggs. **PREDATORS**: Humans, morning sun stars, snails, fishes. **TIDAL ZONE**: Low intertidal to about 300 feet deep. **RANGE**: Southcentral Alaska to Baja California, Mexico. **LIFESPAN**: Up to 35 years.

## Did You See Me? Tell Your Story! _____

_____

_____

**DID YOU KNOW?** *PLEASE*! Don't touch or collect sea stars! We have enough predators in the ocean! So many of us died from Sea Star Wasting Disease. Warming oceans and people are our *most* threatening predators. Help us! Don't hurt us!

42   *Leather Star*

Leather Star    43

# Daisy Brittle Star

*(Can you find a Native name for me?)* • *Ophiopholis aculeata*

*You* haven't heard of a *brittle* star? Well, I'm Betty Brittle, a daisy brittle, the most common of all brittle stars. See my five long, thin arms? They are *so* flexible, they almost look like free-floating streamers! If you or an enemy touches one of these arms, it will drop right off. I *am* brittle, but lucky for me, my arms regenerate, regrowing rather rapidly.

My twisty arms are *so* unique, *nothing* like a regular star's arms. The top side is just one row of tiny, hard, oval plates, held together by joints and surrounded by little plates. Six or seven stout, sharp spines stick out from the side of each center plate. Those stiff spines? They are my pointed tube feet! I don't have round, suctioned feet like other stars, and I don't crawl or walk on my tube feet. In fact, I don't walk at all. I simply stretch out my long, flexible arms and pull myself along the ocean floor. I slither and slink, wriggle and waggle, moving like a stealthy snake. Yes, I'm called the Snake Star!

I'm not *really* a snake, not with my wiggly arms growing from a circular, star-shaped, central disc. My beautiful, small, flat disc extends out and over my arms. It has five rounded points that fall in between my arms. And like them, my disc is made of large, oval plates covered with fine, blunt spines. If you see my underside, you'll see my mouth surrounded by five, movable jaws full of teeth ready to chew anything I grab!

I chew the little bits of food my tiny tube feet help me find. They "sense" when a morsel of food is floating by in the water or lying on the ocean floor where I'm slithering. Those smart little feet reach out, grab that tidbit of food, and pop it right into my jaws. If I'm lucky, a bit of food brushes up against one of my arms and sticks to the gooey mucus all over it. Such an easy snack! I eat just tiny bits of food, so I *never* have to throw up my stomach to feed like other stars. That's such a messy way to eat!

I wonder. Have you guessed I'm very shy? I hide all day and only look for food at night. I stay under rocks, in crevices or abandoned snail shells. I even burrow into debris from broken up shells and dead kelp. I'm so shy you'll probably only see one of my twisted, color-banded arms waving out from under my favorite rock. Wave back, but please, don't try to touch me. I'm fragile, and I *will* fall apart.

## My Facts

**SIZE**: Disc diameter: up to 0.9 inches. Arms: extend up to 5 inches. **COLOR**: Aboral side: may be solid color; usually includes red stripes or blotches mixed with brown, yellowish brown, or orange; some purplish. Oral side: whitish. Arms: most have distinct bands. **FOOD**: Carnivorous. Mainly decaying matter, microscopic organisms, some polychaete worms, small crustaceans. **PREDATORS**: Humans, fishes, harlequin ducks, other birds. **TIDAL ZONE**: Low intertidal to about 5,500 feet deep; generally less than 1,000 feet deep. **RANGE**: Circumpolar; Arctic Ocean into N. Atlantic and N. Pacific; Bering Sea to S. California; Japan. **LIFESPAN**: May be unlimited due to regeneration and reproduction.

**Did You See Me? Tell Your Story!** _____

**DID YOU KNOW?** Now, pay attention! *Brittle* stars move more easily and quickly than all other echinoderms. *Daisy* brittles stars? We are one of the fastest of *all* the brittle stars. Plus, there are more brittle stars than *any* other sea stars. Got it?

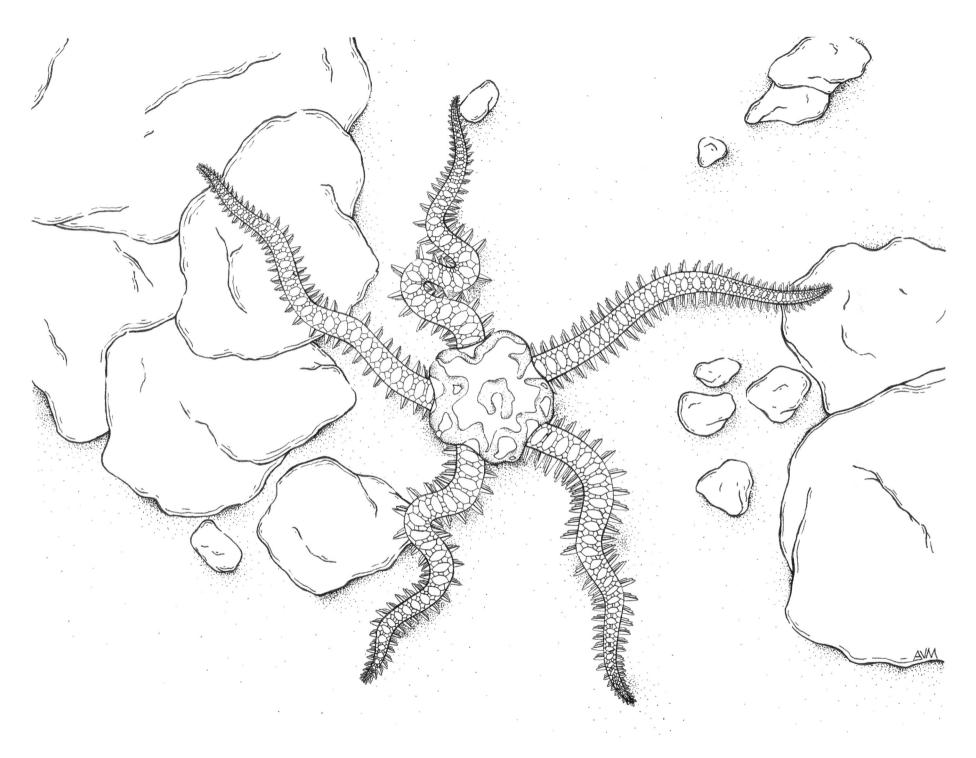

# Red Sea Urchin

*staw xasaa* in Haida • *Mesocentrotus franciscanus*

*What* do *you* want? *Yeah*, I'm Urvil Urchin, and I *know* I'm a red sea urchin. I'm called the porcupine of the sea. So *what*? I have spiked skin, and I *am* prickly! *Yes*! I'm the old man of the sea at a hundred eighty-four years. I *know* that's longer than any land animal and most marine animals live. Age *doesn't* matter. I fix or regrow damaged body parts. Missing ones, too. *Yes*! I *am* as healthy as I was a hundred years ago! And, *yes*, I'll keep on growing, slowly, unless some sea otter or *your* kind kills me. Reds? We're the oldest *and* largest urchin living along this west coast. Did *you* know that?

And do *you* know what my test is? No? *Ha*! It's my inner skeleton, the case that protects my guts and innards. It's a globe-like thing with ten, bone-hard plates. They're stuck together tight, like the sections of that juicy *orange* fruit you eat. My long, thin, flexy tube feet poke out through holes in every other section. Those suction cups on the end of my feet? They help me get away from predators or find food and grab it.

Bet you won't get *this* one! See the hundreds of long, sharp spines on my test? What makes 'em stick? Give up? *Ha*! It's the round openings in bumps all over my test. The non-pointy end of each spine has a ball that fits into those round openings. Skin and muscles cover my test, so I flex my muscles and the hollow spines move in all directions. They're stilts, too, that work with my tube feet to push and lift my test all over the ocean floor. I graze on marine plants when I move, clearing everything around me. I move like I'm *your* age, going a big twenty inches a day.

*Alright* already. *One* more question. What's my Aristotle's lantern? It's my complicated mouth. Aristotle was an ancient Greek scientist, an old guy who wrote about my teeth. He said that my mouth, on the underside of my test, has five, rock-hard "teeth." They point to the inside, and he said it looked like an old-fashioned, five-sided lantern. *Now* your kind says it's a five-sided bird's beak. *Whatever.* This "lantern" is all I need for scraping algae off rocks, grasping and grinding seaweed, and pulling food off the seafloor. Tube feet help hold food while my teeth chomp. Rocks wear my teeth down? Yeah, but that's no big deal. I just grow new ones. No false teeth for this "young" man!

## My Facts

**SIZE**: Test diameter: up to 7 inches. Height: up to 2 inches. Spines: up to 3 inches long. **COLOR**: Test: light to dark purple. Spines: red, dark burgundy, dark purple, brown. Tube feet: dark red. **FOOD**: Mainly herbivorous. Seaweeds/kelps, algae, diatoms; some small barnacles, other invertebrates. **PREDATORS**: Humans, sunflower sea stars, leather stars, other stars, sea otters, crabs, wolf eels. **TIDAL ZONE**: Low intertidal to about 300 feet deep. **RANGE**: Kodiak Island, Alaska, to Baja California, Mexico; N. Japan. **LIFESPAN**: up to 200+ years.

## Did You See Me? Tell Your Story! _____

_____

_____

**DID YOU KNOW?** I stay in rocky tide pools or on rocky shores where waves don't crash. Young urchins, small fishes, and tiny invertebrates hide out in my spines. My spines will hurt *you* if you press hard on them. Leave. Me. Alone. *Understand?*

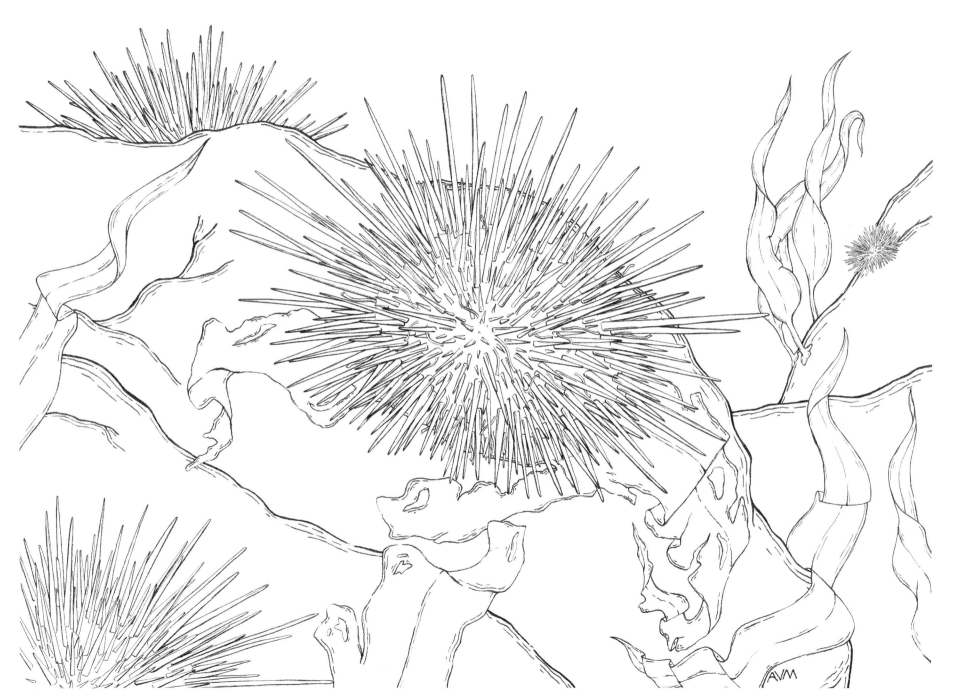

Red Sea Urchin   47

# Green Sea Urchin

*dghezha* in Kenai Tanaina • *Strongylocentrotus droebachiensis*

Ursie Urchin here, a small, green sea urchin with one of the longest scientific names of any animal. *Strongylocentrotus droebachiensis*. I can't even say it, can you? It's too big a name for my slightly flattened, globe-like test that's less than three inches across and an inch-and-a-half high. And the short, thin spines that completely cover my test? They're no more than an inch long. I'm small! But I'm the only echinoderm who lives in the cold water of three oceans . . . the North Pacific, Arctic, and North Atlantic. So there!

Anywhere I am, I have a huge impact on the rocky, subtidal environment around me! You see, I *absolutely* love seaweeds! In fact, I'm green *because* I eat so much green algae, kelps, and other seaweeds. I usually graze with other urchins, staying in one area until all the seaweeds are gone. We "clear cut" the "trees" in that marine forest, creating an "urchin barrier." We destroy the food and hiding places for crabs, sea stars, fishes, and other marine animals living there. They *all* have to move. See how even little critters like me change the balance of a shoreline community?

I'm hard to spot. At night, check tide pools at low tide, near-shore kelp beds, or sandy ocean floors. That's when I'm out eating kelp, seaweeds, and algae. I use the five, hard teeth of my Aristotle's lantern to feed, tearing seaweeds and easily scraping algae off rocks where it grows. (Put me in an aquarium, and my strong teeth will scratch the inside walls of the tank if algae is growing there!) During the day, though, I hide. I tuck away in rocky crevices or caves, on the underside of rocks, or even in a hole some creature dug into a rock. I only leave my safe hiding place at night, then hole up during the day.

Another reason I stay out of sight is the sun. It dries me out! But I have a special cover-up trick. Hidden in my spines are tiny, thin pinching jaws, each on its own flexible stalk. When the stalks stretch taller than my spines, my jaws grab tiny bits of seaweeds, small rocks, or shells. Then my long, narrow, flexible tube feet pick these up and hold on to them with their suctioned tips. I have rows and rows of feet, from my top, aboral center all the way down the sides of my test to my mouth. Feet everywhere! That means tiny pieces cover *all* of me, keeping me hidden and protected from predators *and* the sun! So clever!

## My Facts

**SIZE**: Test diameter: up to 3 inches. Height: up to 1.5 inches. Spines: up to 1 inch long. **COLOR**: Test: greenish with reddish brown. Spines: pale green. Tube feet: brown to purple. **FOOD**: Mainly herbivorous. Seaweeds, kelps, algae; some small fishes and invertebrates. **PREDATORS**: Humans, sea stars, sea anemones, hairy tritons, king crabs, wolf eels, otters, ducks, crows, ravens, gulls. **TIDAL ZONE**: Low intertidal to about 3,700 feet deep. **RANGE**: Alaska to Puget Sound, Washington; North Pacific, Arctic, Atlantic Oceans. **LIFESPAN**: 4 - 8 years.

## Did You See Me? Tell Your Story!

_____

_____

_____

**DID YOU KNOW?** I grow half an inch each year. My test enlarges by adding more and larger hard plates. Touch me lightly, and the spines attached to my test move *toward* you. My short, sharp spines won't poke you if you're gentle.

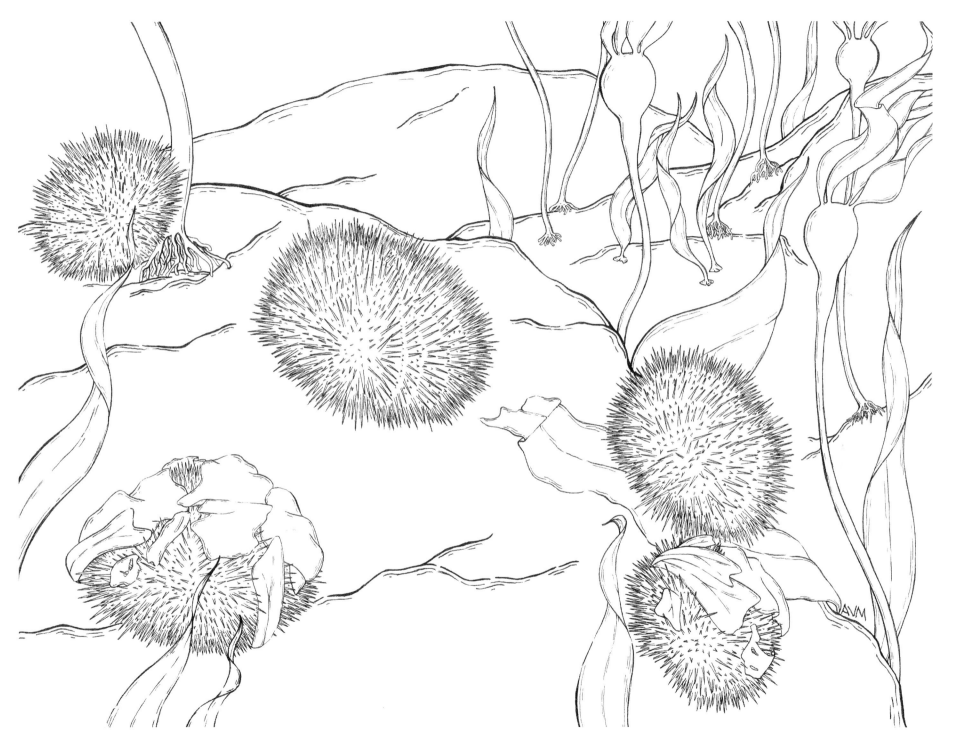

Green Sea Urchin

# California Sea Cucumber
*anaqütak* in Aleut • *Parastichopus californicus*

Surprise! I, Cubby Cucumber, the largest sea cucumber along the northern Pacific Coast, am an echinoderm! I don't look like sea stars, urchins, or the others? No rays, arms, spines, or pinching jaws? I lie on my side, not on an aboral or oral surface? You're right! My body is one long, soft cylinder that looks like a chubby cucumber, an overstuffed sausage. I am *unique*.

Instead of spines and pinching jaws, I have large, pointed, cone-shaped papillae (puh-PIL-ee), or bumps, all over my dorsal side. I *do* have tube feet to help me move . . . five rows of them along the entire length of my ventral, or underside. My endoskeleton, made of tiny, hard, separate skeletal pieces called ossicles (AH-sic-kuhls), is *in* my thick, leathery skin. Perfect! My loose skeleton lets me easily escape a predator. I just relax the five, long muscle bands under my skin, becoming totally flexible and bendy. I ooze through small openings between rocks or crevices, openings so small no stars or other enemies even *begin* to fit. Once through and safe, my muscles tighten to my usual shape. Easy-peasy.

My muscles help me escape in other ways, too. I have *moves*, like wriggling, writhing, twisting, and turning to move to safety. If those don't work, look out! My mouth is on the front end of my body, and at the other end? That's my anus, where I poop. Well, when I'm scared, I tighten my muscles, push a powerful stream of water out my anus, totally hosing my enemy as I escape. I might shoot out sticky threads that trip an enemy. If I'm *really* scared, I poop out some of my own inner parts. The enemy stops to eat these, and I keep moving. Luckily, my innards regrow quickly.

I usually eat on the move, using the mop of long hair around my mouth to trap bits of food. That mop's actually twenty or so sticky, branching tentacles. I drag my sticky hair along the ocean floor, sifting nutrients from the sand and mud as I travel up to thirteen feet a day. I might just stay still in the water, waving my tentacles in the current to catch food particles that float by. Loaded with food, my mop retracts into my mouth, and I strip off the food.

So there! I may be ignored, often even overlooked. But *I* know I'm unique and special!

## My Facts

**SIZE**: Length: up to 20 inches when relaxed. Width: up to 2 inches. **COLOR**: Body: dark red, reddish brown, yellowish. Papillae: paler tipped with red. **FOOD**: Mainly omnivorous. Tiny particles of algae, aquatic animals, waste materials of any kind. **PREDATORS**: Humans, morning sun and sunflower stars, other stars, sea otters. **TIDAL ZONE**: Low intertidal to about 820 feet deep. **RANGE**: Bering Sea and Aleutian Islands, Alaska, to Baja California, Mexico. **LIFESPAN**: 5 - 10 years.

## Did You See Me? Tell Your Story! _____

_____

_____

_____

**DID YOU KNOW?** I'm usually in protected water with a slightly moving current. Look on mud, sand, gravel, rocky rubble, or solid bedrock. *Please*! Don't tell divers where I am. They grab me, slice up my muscles and body wall, then *eat* me!

# Hairy Triton

*ts'esx'w* in Tlingit • *Fusitriton oregonensis*

"What's a triton?" you ask. Well, I'm Trina Triton, a large, hairy snail who lives in the ocean. The small snail in your yard and the slimy slug eating your plants? They're my relatives. We're all gastropods (GAS-truh-pods), a huge, varied class of thousands and thousands of snails and slugs. We come in all sizes and shapes, some with shells and some without. We live in salty oceans and seas, in fresh water, and on land. Gastropods are *everywhere*!

I'm the largest sea snail in the intertidal zone along the entire northwest coast. And I'm pretty special! The state of Oregon chose me to be their official state snail, naming me the Oregon Hairy Triton. It's such an honor to represent my class since there are so many of us, and we're all so different.

Yet our bodies have the same parts: a head, foot, guts, and a sack-like mantle to hold and protect our innards. My mantle, like all shelled gastropods, oozes out a special gel that hardens into my single shell. At first, tiny me fit into its apex, or blunt top point. As I grew, new whorls, horizontal sections, "grew" to make room for my bigger body. Now, my large, spiraling shell is shaped like a chubby, long cone. It's thick and hard, covered with hair, and great protection for my soft body inside.

On my head, I have two pairs of tentacles. The shorter, front pair smells, feels, and tastes what's near. The pair further back on my head is longer and has an eye in each tip that sees light and shadows. My mouth, on the lower front of my head, has a soft, flexible, belt-like radula (RAJ-oo-luh), or tongue. It's covered with hundreds of tiny, spiny teeth in neat little rows. These teeth are so sharp they drill holes right through the shells of my prey. Then the radula pulls the soft, inner parts out of the hole into my biting jaws that grind the food into tiny pieces. An easy meal!

I hunt for food by pushing my head and strong, flat, muscular foot out of the long, large opening of my shell. I glide stealthily on my foot along rocks and the ocean floor, tentacles alert to detect prey. If *I* become the hunted, I immediately pull my head then my foot inside. All tucked in, the thick, leathery trapdoor attached to my foot, called an operculum (owe-PUR-kyuh-luhm), completely seals the opening. Safe, one more time!

## My Facts

**SIZE**: Shell length: up to 6 inches. Shell width: up to 3 inches. **COLOR**: Shell: outside, light brown covered with gray-brown bristles; opening, white; operculum, brown. Body: pinkish yellow mottled with maroon or black. **FOOD**: Carnivorous. Sea urchins, clams and other bivalves, sea squirts, sea stars, brittle stars, polychaetes. **PREDATORS**: Sea stars. **TIDAL ZONE**: Very low intertidal to +/- 500 feet deep. **RANGE**: Bering Sea to S. California; N. Japan. **LIFESPAN**: Unknown.

## Did You See Me? Tell Your Story! _____

_____

_____

**DID YOU KNOW?** Small, raised ribs make a checkerboard pattern on the whorls of my shell. Bristly hairs cover my shell so you won't see the pattern. If you see my shell washed up on a beach, you'll know I've died. Bet a hermit crab moves in!

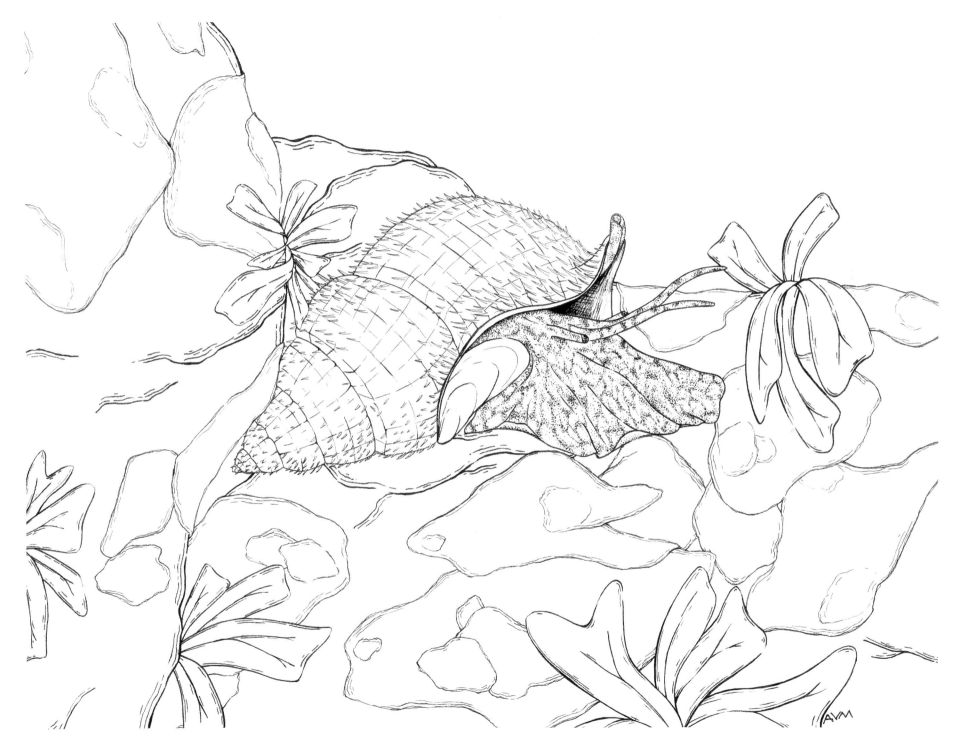

Hairy Triton 53

# Sitka Periwinkle

*gganalggi* in Kenai Tanaina • *Littorina sitkana*

Hey! Down here on the dark rock in the shady part of this tide pool. It's me, Philip Periwinkle, a small, squatty, globe-shaped sea snail. My shell's dark, so I stay in dark places to hide. Color me dark brownish black if you want. See the bands going around my shell from top to bottom? They show the ridges and trenches of my four, spiraling whorls. Color those orangish yellow if you feel like it. On top, my apex is roundy instead of pointy like some snails. It's black. Unfortunately, you won't see my large, completely rounded, pearly-white shell opening. My big foot's sticking out that opening right now, gluing me to this rock.

See the other three periwinkles in this pool? They're different colors. Petra over there, she's black with yellow bands on a smooth shell. Portia, she's pretty common looking with her boring gray shell. The bands on her obvious ridges are dark then almost white then dark. Paulie's on a lighter-colored rock, and his shell is light brown with dark bands. Our shells are the same shape but have different colors. Our fleshy bodies inside our mantles are all black with black antennae.

You probably won't see my body. I'm not *exactly* active, moving less than three feet a month. My pudgy foot has a line down the middle, and when I *do* move, I only lift half of it at a time, and very slowly. You won't see me eat, either. I only do *that* every two to three weeks. When I *am* hungry, I use the sharp teeth on my tongue-like radula (RAJ-oo-luh) to slowly scrape algae, diatoms, and tiny barnacles off rocks and right into my mouth.

Living in a calm tide pool works for me unless the tide comes up too high for too long. I'm one of few invertebrates who breathe real air. I suffocate if I'm underwater very long. But I stay out of the water for several days when the tide isn't high enough in my pool. I don't dry out because I trap some moisture inside my shell and close the opening with the thick, spiny flap on my foot. Glue seeps out of my foot to seal that flap good and tight. Air is in, and heat's out.

Other than pools, look for us in rocky places, crevices, and on pilings if the water's calm. Some of us may hang out in more exposed places without many waves. Remember: we come in lots of colors. Look for a short, squatty, globe-on-a-rock.

## My Facts

**SIZE**: Shell length: 0.4 - 0.78 inches. Shell height: about same as length. **COLOR**: See story. **FOOD**: Mainly herbivorous. Diatoms, algae, black lichen, tiny barnacles. **PREDATORS**: Sea snails, sea stars, red rock crabs, fishes, worms, ducks, shorebirds. **TIDAL ZONE**: Splash or very high tidal to low intertidal. **RANGE**: Bering Sea to coast of Oregon; coastlines of N. Japan and Siberia. **LIFESPAN**: About 2 years.

## Did You See Me? Tell Your Story! _____
_____
_____

**DID YOU KNOW?** Our females lay from 50 to 400 eggs, all wrapped in a thick, jelly-like mass. The mass hangs from the underside of rocks or seaweed. After a week, the larvae begin growing shells. In about a month, tiny sea snails hatch.

Sitka Periwinkle 55

# Opalescent Nudibranch
*(Do you know my Native name?)* • *Hermissenda crassicornis*

*Everyone* talks about *me*! They say... " Oh, that Opal-Nova Opalescent Nudibranch (NEW-duh-brank)! She's the most beautiful, graceful animal in the entire ocean!" Or "Mother Nature went all out giving Opal-Nova her marvelous colors and form." Well, *yes*, my colors are unique! I'm translucent white, with a brilliant, yellowish-orange line down the center of my back. Shocking, bluish-white lines outline the orange. Whitish-blue lines run down my sides, from the front points of my antennae to the tip of my tail. Oh, the clumps of finger-like projections all over my back and sides? These bright brownish-yellow, golden-tipped tentacles are cerata (ser-AH-tuh), an important part of my defense mechanisms.

I need defense! I'm a sea slug, a snail without a shell. My long, straight, soft body has no place to hide. I do have a pair of highly sensitive, whitish-blue tentacles called rhinophores (RYE-nuh-fors) on top of my head. These are my "nose" to help me sniff out both prey and predators. And out in front, I have a pair of long, projecting tentacles. Set into my body at the bases of these tentacles are my eyes that alert me to shadows of nearby enemies as well as possible prey.

My cerata, though, are my first line of defense! Here's how they help. Part of my diet is hydroids and sea anemones. They both have those stinging barbs called nematocysts (neh-MEH-tuh-sists). The barbs don't sting *me*; instead, they pass *through* my stomach and into the tips of my cerata. A predator that annoys me or comes too close will be warned to stay back when I point my cerata forward, like a porcupine points its quills to warn its enemies. If that predator doesn't get the hint and bites me? It gets a mouthful of stinging cells. Some even *die* from the nematocysts' poison!

I'm *definitely* beautiful, but I'm also dangerous *and* aggressive. When I come head-to-head with another of my kind, we *fight*! We lunge and bite, grabbing chunks of tissue from each other. If I come up to the side or tail of another, I usually get the first bite. And for us, first biter usually wins by eating the loser. We are cannibals!

I won't bite *you*, though, so find me. I'll be crawling slowly on rocky shores or on rocks and seaweed in intertidal pools. I like bays and mudflats, and I even hang on pilings and dock floats. I'm *so* worth seeing!

## My Facts

**SIZE**: Length: up to 3.25 inches: average 2 inches. Width: 0.25 - 0.375 inches. **COLOR**: See story. **FOOD**: Carnivorous. Hydroids, sea anemones, sponges, corals, annelids, tiny clams, each other. **PREDATORS**: Other nudibranchs, fishes, sea spiders, crabs, sea stars. **TIDAL ZONE**: Low intertidal to 120 feet deep. **RANGE**: Kodiak Island, Alaska, to Baja California, Mexico. **LIFESPAN**: Up to 1 year.

## Did You See Me? Tell Your Story! _____
_____
_____

**DID YOU KNOW?** I had a shell when I first hatched from my egg case. As a veliger (VEL-uh-jer), or larva, I settled to the ocean floor and grew. In a few weeks, my shell fell off, and I became a miniature adult with all my beautiful colors.

# White Cap Limpet

*sawak'iitaq* in Kodiak Alutiiq • *Acmaea mitra*

Say "Hello" to one of the most stubborn creatures in the entire ocean. Lillian Limpet's the name, a soft-bodied sea snail who lives inside a thick shell shaped like a mini-volcano. Yeah, I'm the Chinese-Hat limpet, and the Dunce-Hat one, too. (More about that later!) Those names describe my shell, the tallest, most cone-shaped one of all limpets living along the west coast of North America. There's no hole in my apex, or top, but my shell's wide, round base is totally open. That means my soft body inside is completely exposed to predators. And *that's* where my stubborn comes in.

When an enemy comes near, I yank the wide, round base of my shell down hard, totally tight around my body. My big, round, muscular foot clings to the rock I'm on, holding on like the world's biggest suction cup. Thick mucus secreted from my foot hardens, gluing me so tightly to the rock *no one* pries me off. That enemy would have to break my shell into pieces to grab my little body! Plus, with my edges completely sealed, direct sunlight during low tide won't dry me out. Heavy wave action? Bring it on!

*I'm* able to unglue and loosen my grip when I need to breathe and to graze for food. But then, I'm open to attack from predators! I'm the only limpet who doesn't "sense" when an enemy is around, so I have *no* defense. Well, I solved that one! My favorite food is pink coralline algae, a tiny marine plant that grows on the rocks and other hard surfaces where I live. I spend so much time eating this plant, it actually *grows* on me. My usual white shell is covered with knobby, pink clumps of the stuff. A simple camouflage and great protection!

What's not so simple is something I *have*, and your scientists *want*. They say my teeth are made of the strongest natural material on Earth, stronger than spiders' silk. You know that gastropods have a tongue-like radula, right? Mine is covered with about one hundred rows of tiny teeth. When I scrape tasty pink algae off rocks, I only use the teeth on the outside ten rows. These teeth wear out in about forty-eight hours and are pushed out by the next rows of teeth. I keep growing teeth that begin in the back of my mouth and slowly move to the front as they grow. Well, your scientists keep trying to figure out what makes my teeth *so* incredibly hard. *HA*! I won't tell. I'm *stubborn*!

## My Facts

**SIZE**: Base diameter: to 1.5 inches. Height: to 1 inch.
**COLOR**: Shell: inside, dull white; outside, white tinged with pink/red algae. **FOOD**: Herbivorous. Pink/red coralline algae.
**PREDATORS**: Sea stars, crabs, fishes, mammals, shorebirds, including black oystercatchers, scoters. **TIDAL ZONE**: Low splash zone and intertidal to shallow subtidal to 100 feet deep.
**RANGE**: Aleutian Islands, Alaska, to Baja California, Mexico.
**LIFESPAN**: On bare rocks, up to 16 years.

## Did You See Me? Tell Your Story! _____
_____
_____

**DID YOU KNOW?** My scientific name means "pointed cap," which led to other names. I'm the "Chinese-Hat" limpet named for an Asian straw hat. "Dunce-Hat" limpet is an old-time school name. Do you know what a dunce hat is? Ask Dad!

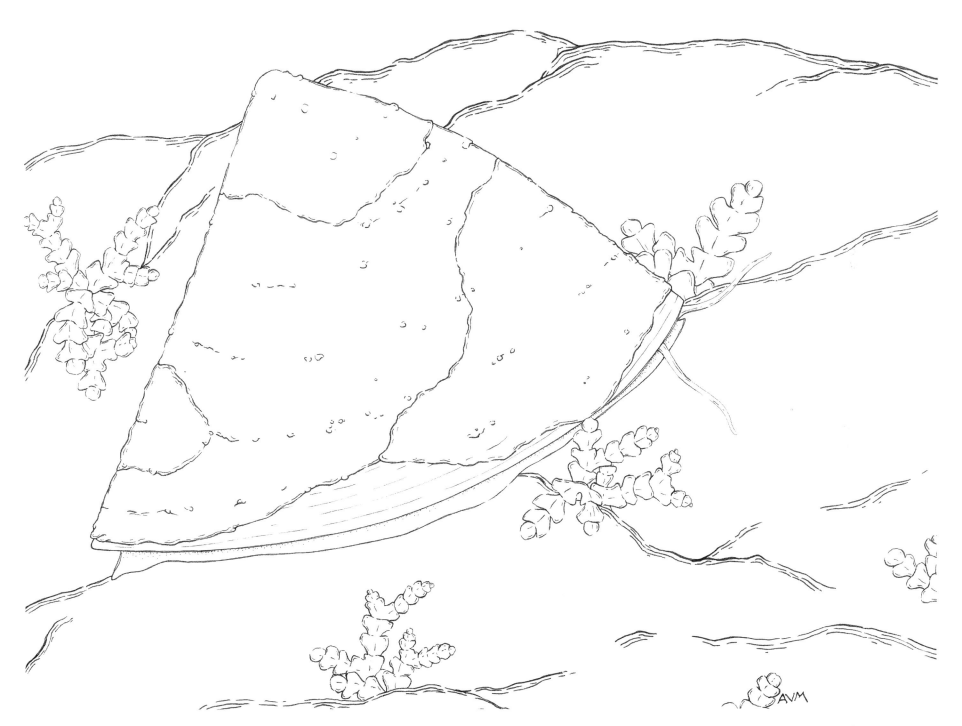

White Cap Limpet 59

# Giant Pacific Octopus

*amiguk* in Dena'ina Qenaga • *Enteroctopus dofleini*

I, Sir Oliver Octopus, am the smartest, most amazing monster of the deep. I'm a *giant* Pacific octopus, the world's largest octopus. The front of my rounded body is my small, broad head that holds my incredibly large brain, two round, bulging eyes, and my mouth. In my mouth are my sharp, hard, parrot-like beak and razor-sharp radula. From the base of my head hang eight, equally-long arms that are my one foot divided into equal parts. The rest of my large, bulbous body? A soft, muscular mantle, the protective case holding my three hearts, ink sac, and other internal organs. Covering me is wrinkly, folded skin with magical cells making me a master of disguise.

I change colors to match my mood and surroundings, becoming the colors and patterns of rocks and ground around me. No shell protects me, so blending with my environs is critical. I hide, too, in the cloudy shape I create by squirting ink. This confuses my enemy, allowing me to *zoom* away. I pull water into my mantle through my two large gills, whoosh it through my body, then shoot it out my siphon tube. I stretch long and thin, so my body squeezes into or out of spaces only slightly larger than my small beak, the only hard part of my body.

Yes, I'm incredibly agile as well as the most intelligent invertebrate on Earth. Your marine biologists capture and study me while I "play" with toys, glide through mazes, open jars, and even interact with them. Some keepers I like; others, not so much. They know how I feel. I've let them figure out that each sucker on the underside of my tapered arms "gives" information to my brain. One arm may grasp your arm tightly; its suckers touch and taste to inform my brain of what's there. At the same time, another arm grabs, tastes, then pulls a nearby crab to my mouth to eat. I'm so clever!

My clever life is short, though. After mating with a female, I swim away to die. She goes to her den and lays fifty to one hundred thousand eggs, depending on her size. Eggs hang in strands from the den's roof. Mom cools and cleans the strands, siphoning water over each one until, in about a year, the eggs hatch. Mom keeps watch until all her hatchlings drift away. Then, having not eaten since laying her eggs, starved and exhausted, Mom dies. And we go on.

## My Facts

**SIZE**: Body, including arm span: 10 - 15 feet average; largest, 30 feet. Weight: male, 90 pounds; female, 50 pounds; largest, 125 pounds. **COLOR**: Body/arms: changes with mood/surroundings; usually deep brownish red with fine black lines. Arms' undersides: mottled white. **FOOD**: Carnivorous. Clams, cockles, crabs, abalones, scallops, shrimps, lobsters, fishes, fish eggs, other octopuses. **PREDATORS**: Seals, sea lions, sea otters, minks, halibut, dogfish, diving birds, other large octopuses. **TIDAL ZONE**: Intertidal to 2,500 feet or deeper. **RANGE**: California to Alaska around Pacific Rim to Japan and Korea. **LIFESPAN**: 3 - 6 years.

## Did You See Me? Tell Your Story! _____

_____

_____

**DID YOU KNOW?** I live in a dark den deep in the ocean where the water's cold. My powerful arms move rocks until I have the perfect hideout. Once I'm inside, I pull another rock across the entrance to seal my den. Smart? Frighteningly so.

# Black Katy/Black Leather Chiton

*kasuugi-x* in Aleut • *Katharina tunicata*

Hey! I'm Katy Chiton (KI-tn), named for Lady "Katy" Katharine Douglas of England. She came to America in the early 1800s, found some of my kind, and sent samples of us back to England to be studied. Scientists say chitons are the oldest group in the mollusk phylum, supposedly a big deal. But *I'm* more interested in people recognizing me today and seeing me in oceans all over the world.

Have *you* seen me? I'm one of the larger chitons at about five inches long. We're all shaped like a flattened football, long, oval, and arched in the center. And, we all wear girdles. Yup, we do. Our shells are eight separate, hard plates that overlap like shingles on a roof. The plates fit together pretty tightly, but that girdle hugs these plates so they won't move at all. My unique, thick, black, leathery girdle is actually the outer edge of my black mantle. It comes out under my shell and at least halfway over my top, covering my white plates to make a cool diamond pattern down my back.

I need the plates tight to protect my soft body inside. I don't have a distinct head, nor do I have eyes, arms, nor tentacles. My mouth is on the front, underside of my body. I'm like all chitons and mollusks with my tongue-like radula and sharp teeth that scrape tasty algae off rocks. Chitons have a second, smaller, sensing "tongue" that pokes around "tasting" for food. The largest part of my small body is my flat, powerful, snail-like foot. It clings so tightly to rocks even high waves crashing along open coastlines don't knock me loose. I stick like glue!

Somehow, though, Native people have been prying my big foot loose for more than three thousand years. Known as a "sea prune" or "bidarki," I'm an important food for some coastal, indigenous people. I'm even part of their traditional stories and legends.

I'm also important to the intertidal ecosystem where I live. I'm not sensitive to light like other chitons, so I graze on algae day and night in sun and shade. Often, hundreds of us feed together, and we wipe out huge areas of algae. Less algae? More sunlight in deeper water. More sunlight instead of darkness? More diatoms, a favorite food for limpets. More limpets, or any species? The balance of an ecosystem is changed. Once again, nature teaches us how *everyone* makes a difference on our planet!

## My Facts

**SIZE**: Length: up to 5 inches. Width: up to 2 inches.
**COLOR**: Plates: white, gray, greenish (algae stained). Girdle: black. Underside: peach, dull orange, or yellow. Foot: pinkish red to darker orange. **FOOD**: Omnivorous. Mainly brown and red algae, including kelps, sea lettuce, diatoms; will eat sponges, tiny barnacles, polychaetes, bryozoans. **PREDATORS**: Humans, leather stars, sea urchins, black oystercatchers, glaucous-winged gulls, sea otters.
**TIDAL ZONE**: Middle to low intertidal to 130 feet deep.
**RANGE**: California to Alaska through Aleutian Islands to Kamchatka, Russia. **LIFESPAN**: 3 years.

## Did You See Me? Tell Your Story! _____
_____
_____

**DID YOU KNOW?** I love rocky shores where strong, heavy waves crash. When the tide's *in*, I crawl around on my foot and eat. When the tide's *out*, I cling to one rock until the water returns. I'm a tough chiton, braving waves *and* sun!

Black Katy/Black Leather Chiton   63

# Lined Chiton

*sk'iwdaangaa* in Haida • *Tonicella lineata*

There is absolutely *no* contest! I win *all* prizes for the most colorful, attractive chiton (KI-tn) living along the west coast of North America! I'm Crazy Kip Chiton, a lined chiton sporting the most zigzag lines, straight lines, bands, blotches, and colors *ever* seen on a small, domed, oval-shaped shell. Eight overlapping valves, or plates, make up my shell, and they're surprisingly smooth and shiny. And look at the colors! The plates start out a basic brown or red, bright blue, or even some shade of yellow orange. And *then*, the fun begins!

Each plate has wild pink and blue zigzags, or red or white lines. I might have blue, purple or black lines, some straight, some zigging. My girdle, the outside part of my mantle that keeps my plates from falling apart, is smooth and thin. It covers just the lower edges of my eight plates. And does that girdle add to my crazy colors! It's brown to reddish pink with yellow or white patches. *All* my crazy colors work together perfectly to camouflage me on the rocks where I live and eat. Like Lillian Limpet, first on my menu is that pinkish-red coralline algae. When I hang out on rocks with that algae, my colors totally hide me. I graze slowly, no worries about being seen.

I hold onto those rocky surfaces with my big, muscular foot. I don't stick super-tight, though, so a predator or extra-big wave will knock me off. When that happens, my girdle and hard plates are flexible enough that I just roll up into a ball to protect my soft, edible underparts. Chitons have earned the name "sea cradle" because we look like a cradle when we curl up. Others think we look armored, like an armadillo or knight of long ago. You can bet *their* "shells" aren't as colorful as mine, though!

Their shells don't have tiny light sensors like mine, either. These sensors let me know whether it's day or night. If the tide's in, and it's dark, I let go of my rock to graze. When it's daytime, or the tide's out, I don't eat. I grip my rock tightly and sleep. I'm not thrilled when the tide's out because I need the oxygen in water to breathe. When I'm above the tideline, I relax and barely move so I don't need as much air. Once the tide's in again, gills in the grooves on either side of my foot work extra hard to catch up on the oxygen I need. Somehow, it all works out. Can *you* work out how to color me bright and unique?

## My Facts

**SIZE**: Length: up to 2 inches. Width: up to 1 inch. **COLOR**: See story. **FOOD**: Omnivorous. Coralline algae, including diatoms, bryozoans, crustaceans, etc. that live on upper layer of the algae. **PREDATORS**: Ochre sea stars, other stars, Harlequin ducks, sea otters. **TIDAL ZONE**: Intertidal to subtidal to 300 feet deep. **RANGE**: Aleutian Islands, Alaska, to S. California; Sea of Okhotsk, Russia, and N. Japan. **LIFESPAN**: Up to 16 years.

## Did You See Me? Tell Your Story! _____
_____
_____

**DID YOU KNOW?** When chitons die, our eight plates separate. Our girdles no longer hold them together. Each plate is called a "butterfly plate" because of its shape. If you see one washed up on shore, finders keepers. A prize for you!

Lined Chiton 65

# Pacific Littleneck Clam

*tl'ildaaskeit* in Tlingit • *Leukoma staminea*

*Yó!* Anyone out there ever eat clam chowder? Fried clams? You may be eating *me*! Clem Clam's the name, a Pacific littleneck clam that's been a favorite food of your kind for over a thousand years. Sometimes called steamer clam, rock clam, or hard-shelled clam, there are more of us in more places than any other clam. Along the entire Pacific coast, we're the smallest, tastiest clam harvested commercially *and* by sportspeople . . . a *huge* deal in your clam-food world.

My plump, tasty body is quite unique! You won't find a head or tail . . . I don't have those. My mouth is near my foot, my stomach is higher than my mouth, and my food tube passes through my heart. *Ha*! My large, thick, powerful, hatchet-shaped foot is at one end of my body. When I want to stay put, I open the bottom of my shell and push my foot out to anchor in the muddy or sandy seafloor. The other end of my soft body has a siphon (SIGH-fuhn) with its two large, muscular tubes joined together. I poke the siphon up and pull seawater in through one tube. This tube filters out tiny particles of food *and* draws oxygen from the water. Then I push the used water and waste products out through the other tube. Even when I'm buried in mud or sand, I shove my siphon up high enough to reach water for food and air.

I'm a bivalve, so my soft little body is kept safe inside two, same-sized valves that make my shell. These are joined by a tight hinge that's held together by two, strong muscles. *Oh.* See the bump on each valve next to the hinge at the narrow top of my shell? That bump is called an umbo (UHM-boh) or beak. I lived between those two umbones the first year of my life. Then my valves began adding a ring each year, enlarging my shell to keep my slowly-growing body from being squished. Count my rings! That's my age, just like a tree.

My ovalish, heart-shaped valves are thick and usually chalk colored. Lines called ribs run from the hinge at the top of my valves to the wide, lower edges. Ribs cross my growth-ring ridges creating a checkerboard pattern. Makes me stand out from other bivalves. You'll be even more sure it's me if you feel along the inside, lower edge of my valves. I have teeth! Look closely and you'll see those small pointy ridges! Yup, that's me!

## My Facts

**SIZE**: Length: up to 2.75 inches. Width: up to 2 inches. Height: up to 2.25 inches. **COLOR**: Valves: exterior, yellowish gray, whitish, or gray to rusty brown with darker pattern lines; interior, white. Siphon: tubes whitish with black tips. **FOOD**: Herbivorous. Phytoplankton particles, diatoms, bacteria. **PREDATORS**: Humans, snails, octopuses, sea stars, crabs, sculpins, sea otters. **TIDAL ZONE**: Mid to lower-half of intertidal to 33 feet deep. **RANGE**: Aleutian Islands, Alaska, to S. Baja California, Mexico. **LIFESPAN**: 8 - 16 years.

## Did You See Me? Tell Your Story!

**DID YOU KNOW?** I'm called "littleneck" because I'm the smallest of all hard-shelled clams. I burrow my little neck about three inches beneath the surface of packed mud or sand. I'm often on small beaches in protected areas. Find me!

Pacific Littleneck Clam 67

# Pacific Razor Clam

*imani Q* in Inupiaq • *Siliqua patula*

Bet you can't catch me! I'm Claire Clam, the largest, fastest-digging clam on the entire Pacific west coast. If I hear you or some other danger coming, I'm *gone*! I push my wedge-shaped "digger foot" straight down into the wet, packed sand. The foot changes into a hatchet shape, opens up, and holds on like an anchor. My strong muscles contract to pull me deeper and deeper into the sand. I dig down almost an inch a second, and I don't stop until I'm a foot and a half under the surface. Think *you* can dig *that* fast?

When there's no danger, I'm usually just below the surface on surf-swept beaches along the open ocean. My siphon, made of a breathing and a feeding tube, is barely above the sand until the tide goes out. Then, you'll just see a small dimple, or hole in the sand where I've pulled my siphon back under the ground. I might be burrowing under that hole, but I'll disappear super fast if I hear you coming.

If you *do* somehow find and grab me, be careful! I'm named "razor clam" for a reason. The edges of my two long, narrow valves are razor sharp. These edges don't close; they always gape open just a bit, ready to slice into an enemy. Grab me on the opposite, smooth, hinged side, or it'll be messy. For sure you'll cut your hand and bleed. If you grab my thin, brittle shell too hard, it'll break, and my soft body inside will squish all over you. (I'm pretty gooey when I'm raw.) Even my thin, glossy-brownish layer of "skin" called the periostracum (pear-ee-OS-truh-kuhm) won't keep my shell from cracking if grabbed too tightly. In fact, that skin layer wears away as I move around in the sand and as waves crash on me. Pretty useless. My annual rings don't wash away, though. Count those concentric circles, and you'll know my age.

I must warn you about eating me. Yes, I'm a highly-prized food, rated tops among clam fanciers. (Don't tell Clem Clam . . . he thinks *he's* number one.) My meat is very sweet, rich, and tender whether I'm boiled, sauteed, deep fried, or used in chowder. The problem is *any* shellfish can accumulate dangerous levels of a certain marine poison, which causes Paralytic Shellfish Poisoning. It doesn't hurt *us*, but if we have this condition and *you* eat us, you'll become very sick. Health officials know when we're safe to eat, so ask them. Be *very* careful!

## My Facts

**SIZE**: Length: 6 - 12 inches. **COLOR**: Periostracum: shiny brown to olive green. Valves: exterior, white with faint violet tinge; interior, glossy white tinged with pink or purple. Siphon: white. **FOOD**: Herbivorous. Phytoplankton, including diatoms. **PREDATORS**: Humans, crabs, flat fishes and other fishes, sea otters, gulls, ducks, sandpipers, brown bears. **TIDAL ZONE**: Low intertidal and subtidal to 30 feet deep. **RANGE**: Aleutian Islands, Alaska, to S. California. **LIFESPAN**: 12 - 19 years.

## Did You See Me? Tell Your Story! _____

_____

_____

_____

**DID YOU KNOW?** I release 300,000 to 118,000,000 eggs each spring. Sadly, the survival rate for my eggs is so low about ninety-nine percent die. The few who *do* survive develop shells after about sixteen weeks. Only the strongest survive!

# Pacific Blue Mussel

*satmaayu-n* in Aleut • *Mytilus trossulus*

I'm Murdock Mussel, and I gotta stay still and quiet. Charlie Crow's circling above this rock where I'm anchored with hundreds of my blue buddies. That crow absolutely *loves* to eat us, and he's trying to pull one of us off our rock. If he yanks me loose, he'll hold me in his beak, fly up really high, then drop me over a rock. I'll *crash* on it, my shell will *shatter*, and my soft, gooey body inside? *Eaten*!

Unfortunately, my shiny, blue-black, bivalve shell is thin and cracks easily. I'm shaped like a wedge, somewhat pointed at my narrow front end and rounded at my wider rear end. I sort of look like a long, rounded triangle. My surface is smooth except for the thin, concentric ridges my growth rings make. See those rings in my drawing?

You probably see my beard, too. It's the bundle of strong, hair-like, gooey, byssal (BY-sal) threads my flat, slender foot oozes out. When these threads hit seawater, they harden and become super-sticky. Since my flat foot *isn't* sticky, like those of anemones, chitons, and other inverts, it's my beard that glues me to rocks, pilings, other blue buddies, and anywhere else I want to be.

Most other invertebrates don't live in the massive groups we do, either. Our dense clusters are so vast and still, it looks like we aren't even alive. We *are*! In fact, we're one of the most common, easily recognized of all mollusks in the North Pacific. Of course that's led to all kinds of names, like bay mussel, common mussel, and even foolish mussel. *Foolish*? I don't like it, but it's true. We often use our byssal threads to hook ourselves together in huge masses that become so big and heavy, we *all* fall off our rock into the sea. Foolish us become food for predators!

Leading the pack of enemies is horrible Oleg Ochre, the sea star who brags about being a Keystone species. *We* gave him that title! If we go low in the intertidal zone, he's waiting there to eat us! (We're his favorite food.) We go above the tideline? We risk drying out in the sun. Plus, old Charlie's always flying overhead. But all is not lost. When a certain sea snail tries to make a meal of us, we circle it, spit out some byssal threads, and tether that creature to the nearest rock. It starves to death! Find out who *that* sea snail is!

## My Facts

**SIZE**: Length: up to 4 inches. Height: up to 2 inches.
**COLOR**: Periostracum: shiny dark purple or brown. Valves: exterior, blue black; interior, pearly white or gray. Foot: brown. Siphon: light with black tip. **FOOD**: Mainly herbivorous. Phytoplankton, some detritus. **PREDATORS**: Ochre sea stars, other stars, sea snails, anemones, crabs, shorebirds, crows, gulls, sea otters, humans. **TIDAL ZONE**: High to mid intertidal to 16 feet deep. **RANGE**: Arctic Ocean, Alaska, to N. California; Russia. **LIFESPAN**: 1 - 2 years.

## Did You See Me? Tell Your Story! _____

**DID YOU KNOW?** I filter water through my gills, catching tiny food particles on mucus-covered sheets. The particles move to my food grooves where I slowly digest them. *I'm* slow when I move, pulling myself along with my byssal threads.

Pacific Blue Mussel   71

# Why Do We All Have Three Names?

Hello! Clem Clam here to tell you why we all have three names at the top of our story pages. Ready? On my page, it says . . .

## Pacific Littleneck Clam
*tl'ildaaskeit* in Tlingit • *Leukoma staminea*

The first name is my common or popular name, the one English-speaking people use when they talk about me. That name is **Pacific Littleneck Clam**. The second name listed is my common name in one of the more than twenty Alaska Native languages. I am known as **tl'ildaaskeit** to the people who speak the Tlingit language. A Native name is provided to introduce you to some of these beautiful languages. My name is not capitalized in *this* Native language, but sometimes our names *are* capitalized.

The last name listed is my scientific name, ***Leukoma staminea***. Everything living on our planet has a scientific name created by scientists. These complicated names are used to be absolutely positive about the identity of the lifeform being discussed or studied. Seven classifications, or divisions, make up my long scientific name:

**Kingdom**, **Phylum**, **Class**, **Order**, **Family**, **Genus**, and **Species**

("**K**ids **P**rod **C**arefully **O**n **F**ragile **G**listening **S**hores" helps you remember!)

Kingdom is the largest classification, and each division after that becomes more and more specific. My entire scientific name is below, with a brief explanation of the characteristics or meaning of each classification.

| Classification | Name | Characteristics/meaning |
|---|---|---|
| Kingdom | Animalia | Animal |
| Phylum | Mollusca | Has soft, unsegmented body, and mantle containing cavity for breathing/excreting; most have shell(s) |
| Class | Bivalvia | Has two valves hinged together making one shell |
| Order | Veneroida | Fresh- and saltwater bivalve with thick valves |
| Family | Veneridae | Hard-shelled marine clam - Venus clam |
| Genus | *Leukoma* | Saltwater Venus clam with chalky shell |
| Species | *staminea* | Littleneck clam |

A scientific name is based on a Latin or Greek word. "Animalia" comes from the Latin "animale," and "bivalvia" is the Latin word for "two valves." Complete scientific names are so long, scientists use only the last two categories, genus and species, when they talk about me or any other lifeform. That is why my scientific name in this book gives you only my genus and species, ***Leukoma staminea***. Now you know!

# Some Sea Stars' Suggestions for Tide Pool Etiquette

Planning to visit a rocky coast or sandy beach? You're hoping to find tide pools to search for us and other invertebrates? What fun! And then you're going to an aquarium with a touch tank? Wow! We know you want to be the best visitor possible wherever you find us. So please, read and follow our suggestions to have a safe visit *and* to protect us in our homes.

# So You Want to Be a Marine Biologist?

I love nature, especially the ocean and its creatures.

I'm creative and like to solve problems.

I want to help protect the animals and plants in the ocean and keep our planet healthy.

I'm curious, and I like to study and learn.

I like adventure, exploring, and being outdoors, especially at the ocean.

## Do these statements describe YOU? Then maybe a career in marine biology is for you!

Marine biologists study everything from the tiniest living organisms to the largest mammals in the oceans.

They investigate and study . . .

. . . microscopic animals and plants that are critical food sources, like zooplankton and phytoplankton.

. . . sea plants and algae that provide hiding places and food for many marine creatures.

. . . fungi, a unique kingdom with over 1500 species of living organisms that are not plants nor animals nor bacteria.

. . . invertebrates like sea stars, octopuses, marine worms, anemones, jellies, crabs, clams, and many, many others.

. . . vertebrates, including fishes, reptiles, birds, and marine mammals.

### What, exactly, do marine biologists do?

Being a marine biologist is one of the best jobs in the world. You might work in a laboratory or scuba dive underwater with the animals. You might be part of a team that discovers a new species of plant or animal or helps determine how humans are impacting the ocean. You might help protect sea turtles when they first hatch and make their way to their ocean home or determine if plankton is decreasing in a particular area. No matter what you do, you will meet and work with interesting, inspiring people from all over the world.

### Research

Researchers collect and analyze data. They study how marine organisms behave and interact with their environments. What sounds does a dolphin make? Why is a kelp forest disappearing? How do puffins raise their young? What caused sea star wasting disease? Researchers find out!

### Educate

Educators help people learn what is happening in the marine world. At schools, universities, aquariums, museums, and government parks and agencies, the public is taught what researchers are discovering. What is happening in and to our oceans and marine life? How can we be involved and help?

# Meet Mr. Richard, Aquarium Curator at the Alaska SeaLife Center

Hello, and welcome to the Alaska SeaLife Center. I came to Seward in 1977 to help create the Alaska SeaLife Center. Now, as curator, it's my job to be sure things are running smoothly. I have so many interesting responsibilities, including . . .

* checking to see that we have the fish, squid, crabs, krill, and other food needed to feed our animals.
* reviewing reports, like ones showing that the water systems are flowing efficiently in the aquarium.
* looking throughout the Center to be sure the exhibits and displays are attractive and clean.
* meeting a group of people to teach a special program about a sea otter, octopus, or some other animal or fish.
* sitting in on research discussions, then riding out on a special vessel to help release a shark that's been studied.

Here at the SeaLife Center we take care of marine mammals, birds, invertebrates, and fishes. We have a large team of research scientists who study many different animals, including Steller sea lions, octopuses, sleeper sharks, mussels, and spectacled eiders. We have a great education team working hard to help people of all ages learn more about the marine sciences. They work with teachers and schools as well as the general public. Maybe you'd like to have an overnight field trip to the SLC and spend special time learning about the animals.

We also have a large number of animal caregivers and specialists. They feed and take care of our resident animals as well as injured and baby animals who are brought to us for treatment. Caregivers work with our veterinarians to be sure all our living creatures have the right food, environment, and attention they need to thrive. They also arrange for release of animals that have been rehabilitated. Maybe you'd like to be a volunteer caregiver for one of the unique birds, fishes, mammals, or invertebrates.

## Sound interesting? Here are suggestions for what you can do now to prepare for this exciting career.

Have pets and learn about them. Be responsible for their feeding and care, including cleaning up after them.

Learn to swim, snorkel, and scuba dive. Most marine biologists spend some time under the water, and it can be one of the best parts of working with marine life.

Visit aquariums, and as soon as possible, become a volunteer. Cut up fish to prepare food for the animals, clean tanks, floors, or sinks. Be willing to tackle any task needed.

If possible, set up a small-scale aquarium and other tanks for tropical fish. As you can, provide temporary care for small species from ponds and other habitats.

## What about education?

Marine biologists are lifelong learners. They are continuously reading and studying. Read everything you can. Biology, zoology, all the natural sciences, chemistry, and math are important subjects to learn. Know how to research and find information online. Plan on four years of college, and if you want to do research and learn more about a specific animal or plant, you'll want to go to graduate school.

# Glossary

**aboral** *(AH-buh-ruhl)*: side or surface opposite the mouth on an animal
**oral** *(AWR-uhl)*: side or surface of an animal containing the mouth

**algae** *(AL-jee)*: plural of **alga** *(AL-guh)*; simple, nonflowering aquatic plants; includes seaweeds and many smaller forms

**antennae** *(an-TEN-ee)*: plural of **antenna**; slender, movable projections on the heads of insects

**appendage** *(uh-PEN-dij)*: a small, projecting part of an animal's body, like an antenna

**bryozoan** *(bry-uh-ZOH-uhn)*: microscopic, moss-like invertebrates that live in groups of thousands; food for many larger invertebrates

**carapace** *(CARE-uh-puhs)*: part of the exoskeleton on some arthropods; covers head and thorax but not abdomen

**carnivorous** *(car-NIV-er-uhs)*: eats meat only
**carnivore** *(CAR-nuh-vohr)*: an animal who eats meat only

**carrion** *(CARE-ee-uhn)*: the carcass and remains of any dead animal

**column** *(KOL-uhm)*: an upright tube or cylinder

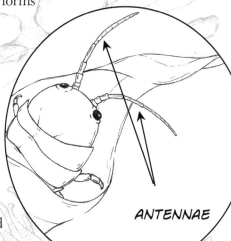

ANTENNAE

BRYOZOAN

**crustaceans** *(kruh-STAY-shunz)*: a class of arthropods, such as crabs, lobsters, shrimps, or barnacles; have hard exoskeletons, antennae, and eyes

**detritus** *(duh-TRY-tuhs)*: waste or debris made from decaying organisms

**diatom** *(DYE-uh-tom)*: a single-celled alga; many species are planktonic

**dorsal side** *(DOR-suhl)*: the back or upper side of an animal
**ventral side** *(VEN-truhl)*: the underside of an animal

**endoskeleton** *(en-doh-SKEL-uh-tn)*: the internal or inside bony plates of an animal
**exoskeleton** *(ek-soh-SKEL-uh-tn)*: a hard outside covering for the body of an animal

**forage** *(FOR-ij)*: to search for food

**gastropods** *(GAS-truh-podz)*: the largest class of mollusks, including snails, slugs, and whelks

**herbivorous** *(er-BIV-er-uhs)*: eats plants only
**herbivore** *(ER-buh-vor)*: a plant-eating animal

**iridescent** *(ear-ih-DEH-snt)*: having a shiny or shimmery color; looks different from different angles

**larvae** *(LAR-vee)*: plural of **larva** *(LAR vuh)*; the immature forms of an animal after eggs hatch

CARAPACE

76   *Glossary*

**mantle** *(MAN-tl):* a fold of the upper or back body wall of a mollusk that holds its internal organs; in mollusks with shells, mantle secretes substance that hardens to form shell

**mottled** *(MAH-tuhld):* marked with spots or smears of color

**mucus** *(MEW-kuhs):* (n) a slimy, slippery fluid
**mucous** *(MEW-kuhs):* (adj) secretes mucus, as mucous membrane

**nudibranch** *(NEW-duh-brank):* brilliantly colored sea slug

**omnivorous** *(om-NIV-er-uhs):* eats both plants and animals
**omnivore** *(OM-nuh-vohr):* an animal who eats both plants and animals

**oral disc**: flattened, upper, round end of the body with animal's mouth in center

**organic** *(or-GAN-ik):* something from natural, living matter with no chemicals added

**pilings** *(PIE-lings):* wooden, concrete, or metal posts pushed into the ground upon which buildings or bridges are built

**plankton** *(PLANK-tn):* small, microscopic organisms drifting or floating in oceans and freshwater lakes; diatoms, protozoans, small crustaceans, and eggs/larval stages of larger animals
**phytoplankton** *(fi-tow-PLANK-tn):* the aquatic plants in plankton
**zooplankton** *(ZO-uh-plank-tn):* microscopic, swimming animals in plankton; usually drift along near surface of oceans and freshwater lakes

**polychaete** *(PAH-lee-keet):* a class of marine worms who live in all oceans; have bristles on their "legs"

**rays** *(rays):* arms on a sea star

**regenerate** *(re-JEN-er-ate):* regrow injured or missing body tissue or part

**rostrum** *(RAH-strum):* a beaklike projection; a stiff snout or front extension of the head in an insect, crustacean, or cetacean

**secrete** *(suh-KREET):* to give off a liquid substance

**sediment** *(SED-uh-ment):* broken down pieces of rocks, minerals, plants, and animals that are moved and deposited in a new place

**tentacle** *(TEN-tuh-kuhl):* a slender, flexible limb or appendage of an animal, often around the mouth of an invertebrate

**thorax** *(THOR-ax):* the chest; area of body between the neck and abdomen

**tunicate** *(TOO-nuh-cut):* a group of marine invertebrates, including the sea squirts and salps, with a hard outer coat and two siphons

**ventral side** *(VEN-truhl):* the underside of an animal
**dorsal side** *(DOR-suhl):* the back or upper side of an animal

**veliger** *(VEL-uh-jer):* a stage in the larval development of some mollusks

**whorl** *(WHOR-ul):* one of the turns of a single-valve shell

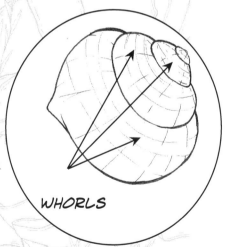

*Glossary* 77

# Our Amazing Oceans

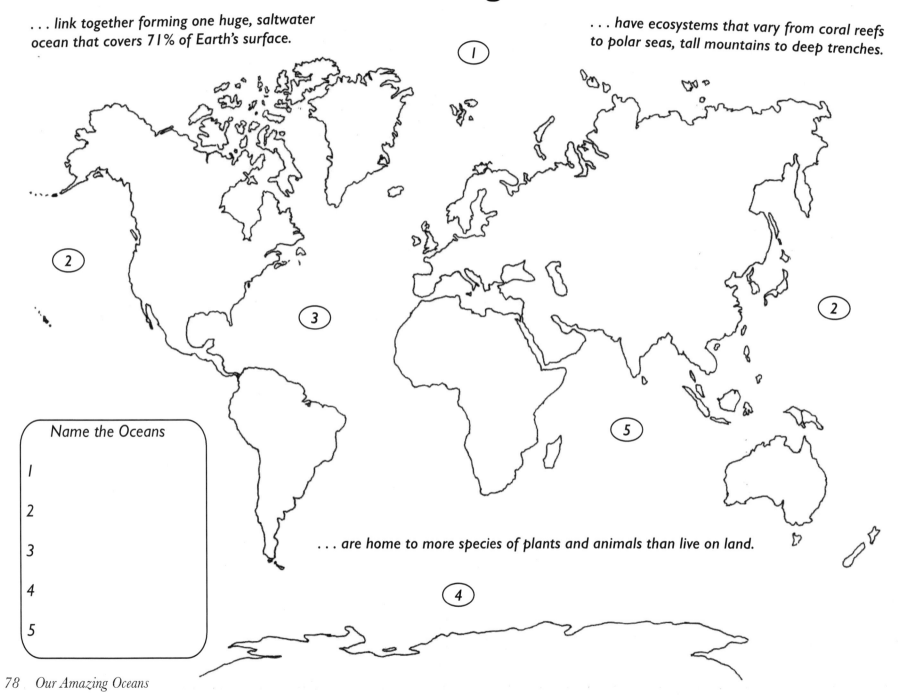

... link together forming one huge, saltwater ocean that covers 71% of Earth's surface.

... have ecosystems that vary from coral reefs to polar seas, tall mountains to deep trenches.

... are home to more species of plants and animals than live on land.

Name the Oceans

1
2
3
4
5

78   Our Amazing Oceans